Pounding Out the Profits

A Century of American Invention

Pounding Out the Profits

A Century of American Invention

* * *

The Development and Manufacture
of Crank-actuated Open-die
Power Forging Hammers

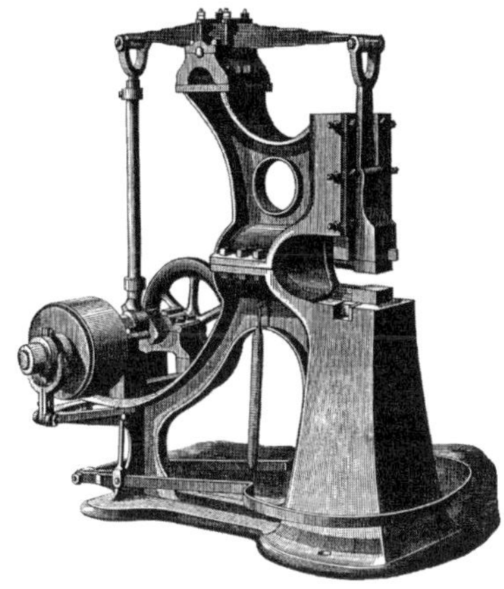

by Douglas Freund

Mingus Mountain Machine Works
Jerome, Arizona

Pounding Out the Profits
A Century of American Invention
by Douglas Freund

First edition - first printing

ISBN 0-9657652-0-2

Published and distributed by:
Mingus Mountain Machine Works
P.O. Box 532
Jerome, AZ 86331

Printed in the United States of America

Contents

Foreword vi

1 *Opening Remarks* 1

"Surely it is Worth While to Study the Hammer Question"

2 *Atmospheric Hammers* 33

"Troublesome Implements" and Ingenious Solutions

3 *Crank-actuated Helve Hammers* 55

"The Favorite of Every Hammerman"

4 *Vertically-configured Guided-ram Hammers* 133

"The Helper with No Bad Habits"

5 *Non-vertically-configured Guided-ram Hammers* 209

"The Hammer that Pounds Out the Profits"

6 *Closing Remarks* 281

"Hooker to Your Ingin"

Notes 287

Bibliography 306

Index 311

Foreword

This book arose from the conviction that knowledge about the history of one's techniques and tools is an important component of craftsmanship.

It was inspired by an extraordinary transformation: machines which had served blacksmiths and manufacturers in a humble and strictly functional manner, are today being used as expressive tools in the hands of artists! An added incentive for an investigation into crank-actuated open-die power forging hammers came from the fact that this class of machine has been largely ignored by historians of technology, and many of the companies which built power hammers are virtually forgotten in the cities and towns in which they once prospered. While much of this neglect is understandable, it is unfortunate. In gathering together the threads of this story, a fascinating picture has emerged. Hopefully, this account of the development of crank-actuated hammers, and the companies which built them, will enhance the remarkable continuing use of these machines.

The author's curiosity about the origins of power forging hammers began while watching artist-blacksmith Albert Paley as he demonstrated his hammer technique at the 1986 conference of the Artist Blacksmiths' Association of North America. Paley was drawing out heavy steel bars into highly-textured organic forms, using a self-contained air-actuated hammer. Surely, this machine was not originally manufactured to be used as an artist's tool. So, what were its intended applications? How many American companies built these hammers? What kinds of hammers were made, and in what numbers? A perusal of nineteenth and early twentieth century engineering magazines produced an ever-increasing list of builders. Summers were spent in the patent library. It became evident that the diverse terms applied to these machines were at least partially responsible for their obscurity. An important step in appreciating the rich history of open-die power forging hammers came with the development of a system of classifications which encompassed and differentiated all machines of this type. It is hoped that others will find this scheme similarly helpful.

While an effort has been made to discuss all American-built crank-actuated hammers in this book, the author is certain that others will come to light. Many of the companies and machines included here became known only through the serendipitous discovery of a rare trade catalogue or an isolated advertisement. The author's research into these machines will continue, and any further information on the hammers discussed here, or word of other hammer builders, will be greatly appreciated. *Pounding Out the Profits - A Century of American Invention*, is a history men and machines of the past. Eventually, a revision is foreseen, filled with illustrations of diverse hammers in use by modern blacksmiths. While crank-actuated hammers form perhaps the largest and richest category of open-die power forging hammers, the history and development of foot-powered hammers (or "Olivers"), as well as steam- and air-actuated machines, are also fascinating. Separate volumes on these subjects are planned for the near future.

* * *

The study of such a multitude of machines and manufacturers was made possible only with the assistance of many people across the country. The on-going correspondence resulting from this research has been, in fact, one of its most rewarding aspects.

In many cases, the companies which manufactured crank-actuated hammers were all but forgotten. Local historians have been gracious in their efforts on my behalf, frequently joining in my enthusiasm and curiosity about these enterprises and their products. I wish to sincerely thank the following individuals and institutions, and acknowledge their contributions to this project: Linda Adkins, *Dexter Public Library*; Shirley Baker, *Butler County Historical Society*; Betty Baule, *Carnegie-Stout Public Library*, Dubuque; Edwin A. Battison, *American Precision Museum*, Windsor; Mary E. Beason, *Powashiek County Historical and Genealogical Society*, Montezuma; Evelyn Bechtel, *New Hampshire Historical Society*, Concord; Joan Bedard, *Worchester Historical Museum*; Philip Bergen, *The Boston Society*; Kurt Brown, *O'Brien County Historical Society*, Pringhar; Sherry Brown, *Massillon Public Library*; Terrence Buckaloo, *Sterling-Rock Falls Historical Society Museum*; Sylvia Bugbee, Shelbourne; Bridget Burke, *The Smithsonian*; Karen Stites Campbell, *Baily/Howe Library*, *University of Vermont*; Tacie Campbell, *Dubuque County Historical Society*; Russell Cashion, Antioch; Emily Clark, *Chicago Historical Society*; Jason Davis, *Danbury Scott-Fanton Museum & Historical Society*; John Dittmeier, Dayton Beach; Eleanor Elwert, *Rutland Historical Society*; Tom Enderson, Jim Falls; Elizabeth H. Ernst, *Grinnell Museum*; Eugene Clarke, Jr., *Chambersburg Engineering Company, Inc*; Linda Evenson, *Freeborn County Historical Society*, Albert Lea; Erin Foley, *Minneapolis Public Library*; Kathryn Furtak, *Schuyler Historical Society*; Leslie A. Gardner, *State Historical Society of Iowa*, Des Moines; William A. Getz, Moline; Michael Gibson, *Center for Dubuque History*; Terry A. Girouard, *Western Museum of Mining & Industry*, Colorado Springs; Grace Gould, *Hennepin History Museum*, Minneapolis; Lorna Baldner Grow, *Dallas County Genealogical Society*, Dallas Center; Mary Henrichs, Albert Lea; Mary Jane Hettinga, *Marathon County Historical Society*, Wausau; Jack Jackson, *Chemung County Historical Society*, Elmira; Velma R. James, Montezuma; Muriel Jeske, *Brown County Historical Society*, New Ulm; Claire D. Johnson, St. Johnsbury; Milo Johnson, Ferndale;

Mary Ann Kane, *Cortland County Historical Society*; Don Kemper, Ridgefield; William H. Ker, Allenwood; Frank Kosmerl, Rochester; Evelyn Kupec, *Joseph T. Ryerson & Son, Inc.*; Paul Lague, *TC/American Monorail, Inc.*, Suellyn Lathrop, *Kansas State Historical Society*; Karen J. Laughlin, *State Historical Society of Iowa*; Peter Leibold, *The Smithsonian*; John Leikvold, *The Challam Co.*, Minneapolis; Elizabeth Lessard, *Manchester Historic Association*; David N. Lewis, *Vigo County Public Library*; Mary Lou Lubinski, *Montgomery County Historical Society*, Dayton; Jim McCarty, Jefferson City; Curt McConnell, Lincoln; John McLellan, Roseville; Emily Miller, *Missouri Historical Society*, St. Louis; John "Ken" Monson, Minneapolis; Maurice J. Montgomery, *Rock County Historical Society*, Janesville; Sonia Moss, *Stanford University*; Gladys C. Murray, *Centre County Library*, Bellefonte; Pat Narver, San Marino; Elinor R. Nielsen, Albert Lea; Kimberly Norris, *Barbour-Stockwell Co., Inc.*; Jane Nowak, *Los Angeles Public Library*; Lavonne Oftedahl, *Albert Lea Public Library*; Thomas Parker, *North Reading Historical & Antiquarian Society*; Marie Panik, *Historic Northampton*; Jo Ann Pell, *Brazil Public Library*; William Pioppi, *Russell-Harrington Cutlery Company*; Jan Rogness, *Albert Lea Public Library*; Grant Sarver, Tacoma; Judith Simonsen, *Milwaukee County Historical Society*; Dick Sitarz, *Prospect Foundry, Inc.*; Francis Skelton, *The Whitney Library*, New Haven; Joyce Smith, *Cortland Machine & Tool, Inc.*; Rick Stager, Birchreenville; H. "Sid" Suedmeier, Nebraska City; Pat Swanson, *Montgomery County Historical Society*, Dayton; Fran Tageson, *Kinney Pioneer Museum*, Mason City; Mark Thomas, San Jose; Page A. Thomas, Fairview; Eldbjorg Tobin, *Chippewa Valley Museum*, Eau Claire; Rudy Turkal, Massillon; Marie Vileta, *Tama County Historical Society*; Margy Vogt, *Massillon Museum*; Margaret Walsh, *Colorado Historical Society*; Lanelle Willig, *Western Iron Works, Inc.*, San Angelo; G. C. Witt, *Bloomer Historical Society*; Frances J. Woodworth, *Red Oak Public Library*; Tom Yeager, Buckeye.

Additionally, I would like to thank the Board of Lick-Wilmerding High School in San Francisco, for a faculty-development grant which enabled me to spend a week at the Smithsonian. Thanks also go to Sherri and David Sacks, and Jaime and Steve Gerigk, for their gracious hospitality while I was on the road; and to Myra, Adam and John for their encouragement. Special thanks go to Tracy for her unwavering enthusiasm, patience, support and editorial contributions; and to my parents, Maxine and John, for their invaluable assistance in proofreading the manuscript. *Pounding Out the Profits* was typeset despite the best efforts of Cleopatra, whose paws entered many unsolicited editorial comments when the author was away from the keyboard.

Douglas Freund, Jerome, June, 1997.

This book is dedicated to my father,

Dr. John Ernst Freund

His passion as an educator and writer, and his delight in craftsmanship, are my inspiration.

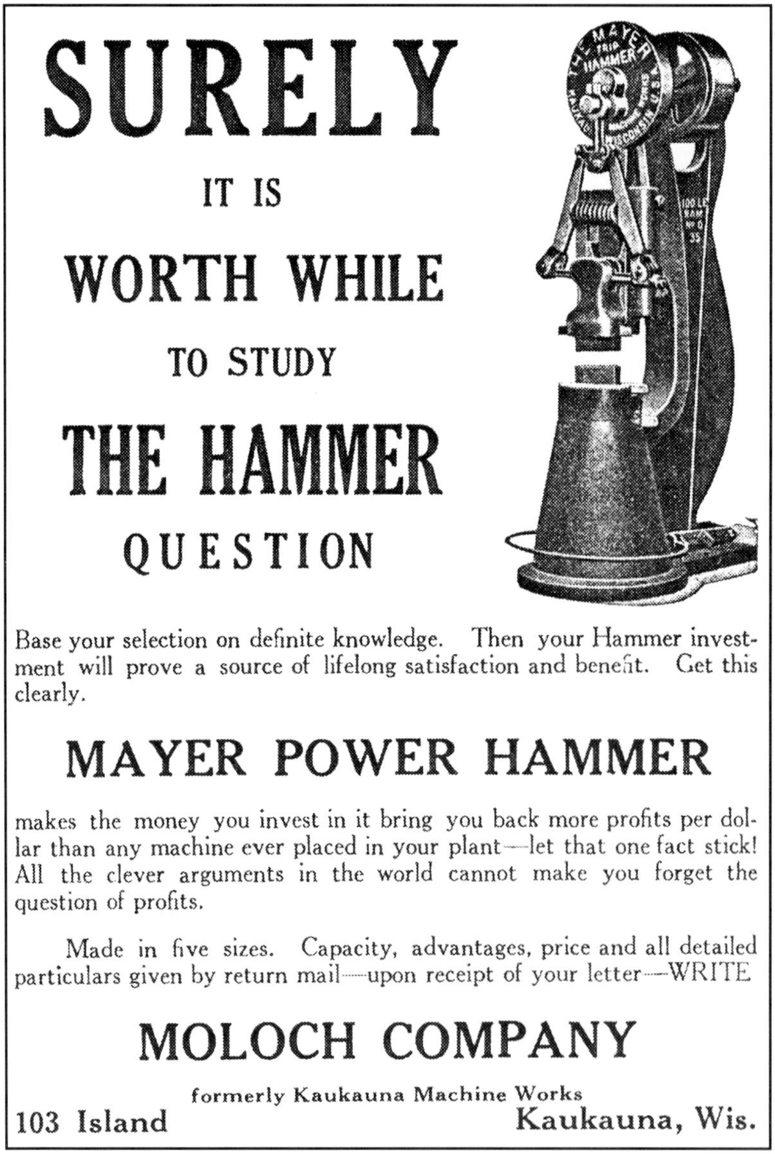

An advertisement from 1921. (*The Blacksmith & Wheelwright*)

CHAPTER 1

"Surely it is Worth While to Study the Hammer Question"

A new generation of men and women now labor at the forge, embracing the blacksmith's craft not only as a profession, but also as a means of artistic exploration and expression. For decades the craft had languished, all but obsolete. This might not have seemed possible at the prosperous dawn of the twentieth century. The blacksmith's profession had grown in stride with the nation: an ever-increasing number of wagon and buggy tires required welding, plowshares needed to be layed, and farriery provided steady bread-and-butter work in many shops. The turn-of-the-century blacksmith was as modern-minded as most Americans, taking advantage of mass-produced goods such as horseshoes and carriage hardware. Affordable gasoline engines made a powered shop possible, rendering small blacksmith's shops more productive and competitive. Motorized forge blowers, grinding stones, drills and power hammers reduced the need for expensive human labor. Power hammers, in particular, diminished much of the drudgery and brute force required in traditional forging operations. Despite these improvements, the blacksmith's trade began to falter. The success of the automobile, ever-increasing mass production, new iron-working technologies, and the exigencies of two world wars eliminated almost all of the smith's traditional niches. Popular appreciation for the blacksmith's work and its potential also began to fade. However, within the blacksmith's craft lay the seeds of its own renewal: the process of forging iron, so demanding of skill in a traditional context, is rich in potential as a sculptural medium. The late twentieth century has witnessed no less than a renaissance of the American blacksmith. Like his or her counterpart from the beginning of the twentieth century, the blacksmith at the dawn of

The state-of-the-art blacksmith's shop at the beginning of the twentieth century was powered by a gasoline engine. In this instance, the power was transmitted by leather belts to a post drill, a grinder, a power hammer and a forge blower. (*The Blacksmith & Wheelwright*)

the twenty-first century uses tools and processes both new and traditional. Power hammers have again assumed an important place in many shops; and interest in the history of these machines has grown. Often the artist-blacksmith relies upon a restored, crank-actuated hammer, built more than half a century ago by a company long since vanished. The transformation of these hammers from simple workhorses to tools used for artistic expression is both remarkable and exciting, but no more so than the rich history of the original development of these machines and the American companies which built them. This book tells the story of those inventors and manufacturers, and attempts to organize rationally the many variations of this prominent type of open-die power forging hammer.

* * *

Crank-actuated hammers make up one of the most diverse and interesting groups of open-die power forging hammers. They appeared at about the time of the Civil War, and were manufactured in the United States for more than a century. They represent a truly American contribution to forging technology, and were exported around the world. Initially, crank-actuated hammers were primarily a manufacturer's tool. They were built in a wide range of sizes, each intended to forge a specific thickness range of iron or steel. When dedicated to repetitive tasks, such as drawing-out the tines of hay forks, preparing file blanks or forging cutlery, they offered enormous savings in cost and labor. They steadily replaced the older trip-actuated machines in America's forges and factories. Crank-actuated hammers were also, to a certain extent, developed initially as an alternative to expensive direct-acting steam hammers. Steam hammers had begun to assume a prominent place in English and French forges during the 1840's. They were soon considered indispensible in shops producing large forgings, such as the shafts for marine engines. A handful of James Nasmyth's steam hammers were imported to the United States, and the rights to manufacture them were assigned to Merrick and Towne of Philadelphia. However, the licensing fee to build such a hammer was so great ($1,800) that a flurry of invention was launched. Among the early American alternatives to the direct-acting steam hammer were so-called "atmospheric" hammers which used an air cylinder as a spring. The vacuum created in the air cylinder could be used to forcefully propel the ram downward, and a cushion of air below the piston could be used to regulate the force of the blow. Ultimately, this technology could not compete with the direct-acting steam hammer in the realm of large forgings. However, there were advantages to a hammer which did not require a boiler, could be run from any factory lineshaft, and could be built to strike rapidly. While some crank-actuated hammers were built with rams weighing up to 1,000 pounds (capable, it was claimed, of forging stock over seven inches thick), they were generally built in relatively smaller sizes. Small direct-acting steam hammers, and later self-contained air-actuated hammers, proved more effective and economical in working stock more than about five inches thick. However, steam hammers, were relatively slow (particularly in the early decades of their development), and crank-actuated hammers gained the advantage as stock thicknesses decreased.

Crank-actuated power forging hammers can be conveniently divided into three categories. First, there are those hammers which employ a helve to carry the hammer ram. These machines were directly descended, mechanically-speaking, from the trip-actuated helves that were used to forge iron in this country from the middle of the seventeenth century (and for centuries before that in Europe). Much like a hand-wielded hammer, the ram of a helve hammer moves in an arc, guided only by the helve. Second, there are the guided-ram hammers where the actuating crank is located directly above the hammer ram. Such vertically-configured hammers have rams which move straight up and down, constrained by guides. Third are the guided-ram hammers which are not vertically-configured. Here, the ram is generally oscillated by a pivoting beam, with the actuating crank located below. The beam may pivot at its end, with the connecting link to the crank (pitman) located between the pivot and the ram, or the pivot may be located in between the pitman and the ram. All three of the types, with their individual virtues and limitations, coexisted and competed with one another for many years. The earliest crank-actuated hammers were non-vertically-configured; however, their design did not include a pivoting beam. These influential machines were of the "atmospheric" type, employing air cylinders rather than metal or rubber springs. For the sake of clarity, they will be discussed first, in a separate chapter.

The companies which built crank-actuated hammers did not necessarily limit themselves to only one of these three general designs. Some builders moved from one type to another in an attempt to better compete in a changing market. For example, the Beaudry Company of Boston, Massachusetts, built non-vertically-configured guided-ram hammers until switching exclusively to vertically-configured machines. Other hammer manufacturers

Principal Types of Crank-actuated Power Hammers

1) A Crank-actuated Helve Hammer

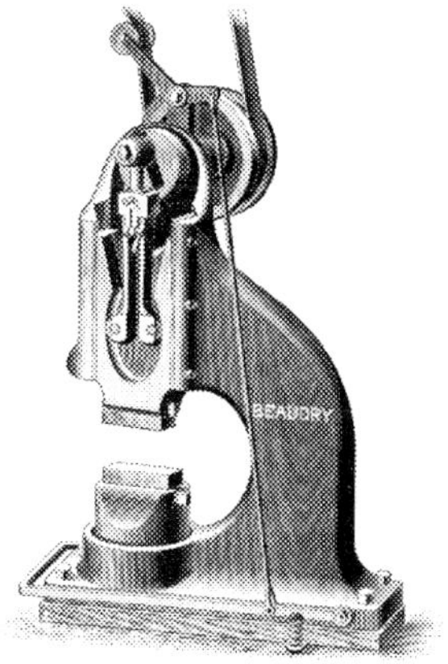

2) A Vertically-configured Guided-ram Hammer

3) Non-vertically-configured Guided-ram Hammers

(The machine at the left has its pivot located at the end of the beam. The hammer in the center has its pivot located between the connections to the crank and the ram. The machine at the right is a so-called "atmospheric" hammer. While it has no pivoting beam, the crank is located below the air-cylinder linkage).

included several designs in an effort to broaden their market. The Mayer Brothers of Mankato, Minnesota, added a helve hammer, the Easy, to accompany their already successful vertically-configured Little Giant. The Bradley Company of Syracuse, New York, developed hammers in all three categories, marketing them concurrently for many years.

The crank-actuated hammers built for manufacturers' use were expensive capital acquisitions. In 1894, for example, a 60-pound Bradley Rubber Cushioned Helve cost $600 and a 50-pound vertically-configured hammer built by the Philip S. Justice Company of Philadelphia, Pennsylvania, cost $380.[1] By the turn of the century, foundries in the Midwest were beginning to offer crank-actuated power hammers at a price which most blacksmiths could afford. By 1909, the Kerrihard Company of Red Oak, Iowa, was selling its hammer for as little as $60.[2] Despite these declining prices, many American blacksmiths devised their own power hammers. The fundamental simplicity of crank actuation made it popular with the blacksmith-mechanic, and many smiths produced workable hammers from old buggy springs, spare pulleys, scrap metal and elbow grease. In a number of cases, hammers of such humble origins were refined and brought to market.

Remarkably little has been written about open-die power forging hammers, even less about crank-actuated machines of this type. Therefore, a number of distinctions and definitions will prove helpful. The term "open-die" refers to the process of forging metal in a relatively unconstrained manner, much like most hand-forging. Open-die power hammers are often fitted with flat-faced dies (which move the metal equally in all directions perpendicular to the force of impact). The use of flat dies further enables the smith to employ an assortment of hand-held tools under the hammer, exploiting the full versatility of the machine. Such hand-held tools may be used to cut the metal, swage it into particular shapes, confine the flow of the metal to specific directions, add texture, etc. For instance, a tool with a cylindrically-curved face, a "fuller," will limit the flow of the metal to directions perpendicular to the axis of the cylinder. Such tooling is used to quickly "draw-out," or

B. F. Talley, of Missouri, stands beside his homemade vertically-configured guided-ram power hammer, 1915. (*The Blacksmith & Wheelwright*)

elongate, a workpiece (the tines of a pitchfork, for example). It is also possible to use a curved upper die, rather than a hand-held tool, to accomplish this same forging operation, but the versatility of the hammer becomes temporarily limited. An open-die power hammer can be fitted with increasingly specific tooling, including so-called "closed dies," but other machines are better suited to such work. Closed dies completely constrain the material being forged (any additional material, or "flash," left at the parting-line of the dies, is trimmed away prior to finishing). Closed-die forgings are typically

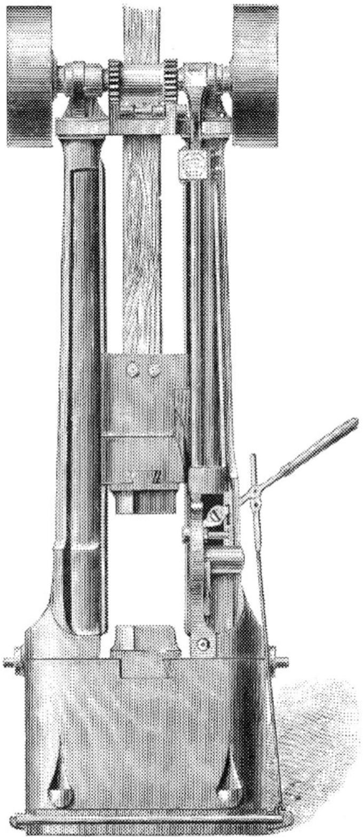

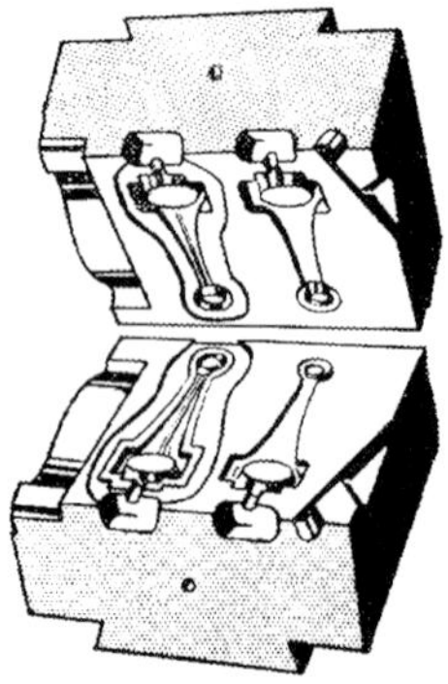

The board drop hammer is an example of a closed-die forging machine. It is capable of striking discreet blows of consistent force. The dies used in such a hammer are engineered to produce complex forgings in a minimal number of blows. The set of dies at the right is intended to produce connecting rod forgings in about ten blows. (Hammer image from *The Masterpieces of the Centennial Exhibition*)

created with a single blow, or a brief succession of blows. For that reason, specialized closed-die forging machines, such as drop hammers, are preferred. Open-die hammers, with the exception of foot-powered machines, are generally designed to provide fairly rapid, continuous blows. They are not designed to provide single blows of consistent force. Foot-powered hammers, or "Olivers," are most like closed-die hammers in providing discreet blows, but they are decidedly unsuited for the rigors of twentieth-century mass-production.

Closed-die forging machines are capable of creating complex forgings in an instant. The cost of the dies may represent a substantial portion of the expense of the forgings, and very little of the traditional blacksmith's skill is required to operate the hammer. In fact, the most sophisticated developments in this field include automation which eliminates the hand and eye of the smith altogether. Open-die power hammers, on the other hand, employ relatively simple tooling, and demand a high level of skill on the part of the hammersmith. While the operation of large open-die hammers often requires a crew, most crank-actuated power forging hammers are of such a scale that they can be operated by a single smith. This is accomplished by means of a foot control. While the operation of the machine is relatively straightforward, the versatility of such a hammer makes highly sophisticated work possible, and readily admits the potential for artistic expression. It is thus very much an extension of the hand-wielded hammer.

Open-die power forging hammers are also to be distinguished from a variety of other highly specialized forging machines such as "upsetters," "roll-forging machines," "rotary

Roll forging is another example of a tooling-intensive process in which precise results are achieved with relatively little skill on the part of the machine operator. The rolls make a partial rotation and reverse; and the work is completed in one or possibly several passes. (*International Library of Technology*)

swaging machines," etc. These machines are ideally suited to the requirements of mass-production, but lack the versatility that so characterizes the open-die hammer. Interestingly, the upsetter accomplishes a fundamental forging operation that cannot generally be performed by an open-die hammer. "Upsetting" refers to the operation of increasing the thickness of a workpiece by decreasing its length (with selective heating at the point to be thickened). With only a couple of rare exceptions, open-die hammers cannot handle a workpiece on-end. Instead, dedicated upsetting machines are used for repetitive tasks such as thickening the ends of rock drills. As with closed-die hammers, the tooling used in such machines is not versatile. Since upsetting is impossible with most open-die power hammers, it becomes necessary to adopt an alternative forging strategy to accomplish the same result: the smith starts out with thicker stock, and draws-out the material which adjoins the wider section. In hand blacksmithing, upsetting is an elegant and efficient process; but it can also be tedious and demanding. The alternative power hammer technique, while it requires far more hammering, is therefore often preferred by the smith.

All of the aforementioned forging machines, including open-die power hammers, exploit the remarkable properties of certain metals, primarily, but not exclusively, ferrous. At elevated temperatures they can be plastically deformed, and upon cooling regain their strength, toughness and resistance to fatigue. Forging improves the grain structure of the metal and orients the flow lines in a manner that can bestow strength at points which might otherwise be weakest. While it might be easier and cheaper to produce a given shape by casting or machining from stock, the strength imparted by the forging process is often a crucial requirement. For that reason, forging has remained a vital industrial process at the close of the twentieth century. However, at the present as in the past, improvements in other materials and processes continue to challenge the place of forging technology. Prior to the development of gas and electric welding processes, iron and steel were welded at the forge. Today such welding is rare. While new, lighter, stronger man-made materials may well replace steel in many engineering applications, the blacksmith and his/her ferrous materials will endure. New composite materials, ceramics and adhesives may eventually prove more practical for mass production, but these materials lack the inherent versatility (and almost endless recyclability) of steel.

Crank-actuated hammers occupy only a part of the full spectrum of open-die hammers. At one extreme are the giant steam-powered hammers, with rams weighing up to 125 tons. At the other extreme are foot-powered hammers, or Olivers, some of

A full-scale model of the giant 80-ton Creusot steam hammer was displayed at the Paris Universal Exposition of 1878. (*Reports of the United States Commissioners to the Paris Universal Exposition*)

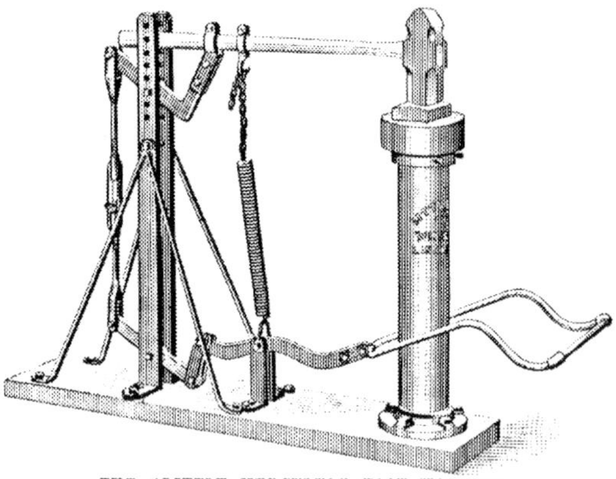

The Little Hercules Oliver was manufactured in Lancaster, Pennsylvania, by Zell & Shenk, ca. 1911. Using a standard sledge hammer (and powered by foot), it was among the least sophisticated of open-die power forging hammers. (*The Blacksmith & Wheelwright*)

which were fitted with ordinary sledge hammers. The largest steam-powered hammers were developed to forge large engine shafts, ordnance and armor plate. By the end of the nineteenth century, these machines had been replaced by hydraulic forging presses. Olivers, on the other hand, were built to provide an alternative to the human striker for hand-forging operations. They continue to be manufactured in the United States, and are often found in the shops of today's artist-blacksmiths. Crank-actuated power forging hammers spanned much of the range between these two mechanical extremes. The smallest crank-actuated hammers were similar in scale to the Oliver, while the largest crank-actuated hammers offered competition to the smaller sizes of steam hammers. Within the size-range of crank-actuated hammers were also trip-actuated machines and self-contained air-powered forging hammers. This latter group has outlived the crank-actuated hammer, and is still manufactured by several companies outside of the United States. Pound for pound, the self-contained air-powered hammer is far more powerful than a crank-actuated machine, and capable of working a greater thickness-range of metal. Such a hammer (new or used) is generally much larger and costlier than a crank-actuated machine. Conversely, crank-actuated hammers with rams in the weight-range from roughly 25 to 200 pounds are often more affordable and manageable for the artist-blacksmith. With replacement parts generally unavailable, restoring a vintage hammer can be a challenge. Recently, though, workshops on restoring hammers such as the Little Giant have become popular and well-attended. There is clearly a potential market for new, small, open-die power hammers. Whether any of these future machines will return to a crank-actuated design remains to be seen.

* * *

To a certain extent it is understandable that so little has been written on the subject of open-die power forging hammers. The only machines of this type which truly captured the public's imagination were direct-acting steam hammers during the nineteenth century. Working steam hammers were often displayed at the great exhibitions of the second half of the nineteenth century, alongside other mechanical wonders. Spectators were particularly impressed by the degree of control with which these hammers could operate. A common demonstration included cracking the shell of a walnut without crushing its meat. On several occasions, full-scale models of the world's largest steam hammers were created, to the awe of engineers and the public alike. For the most part, though, open-die hammers, including crank-actuated machines, worked behind the scenes (if not quietly, at least humbly). They were used to manufacture and repair many objects that Americans touched in their daily lives. But, like most tools of manufacture, they were taken for granted by the public. Historians of technology have also devoted little attention to open-die power forging hammers. This too is understandable. Much has been written about fundamental machine tools such as the lathe and milling machine, but these tools are are still widely used today and continue to evolve. Open-die power hammers, on the other hand, were inexorably replaced in practically all of their technological niches. Perhaps the very diversity of these niches, as well as their extremes of scale, are responsible in part for the fact that open-die power forging hammers have rarely been considered collectively. Such consideration has also been hindered by the imprecise, confusing and often misleading terms applied to these machines.

Crank-actuated power hammers were often referred to erroneously as "trip hammers." Properly speaking, though, trip hammers employ a trip actuation to raise and then release the hammer. However, before embarking upon a crusade to rectify this possible point of confusion, it must be remembered that the term "trip hammer" was widely misused throughout the period in which these machines were manufactured. Inventors, builders, patent office personel, users and the press often referred to crank-actuated hammers as trip hammers. Unfortunately, like so many other successful slang terms, the expression "trip hammer" rolls easily off the tongue, and its use will undoubtedly persist. It can be potentially misleading, however, for during the period when crank-actuated hammers were built, they never completely replaced the true trip-actuated hammers. Trip-actuated hammers were relatively inexpensive to build and sell; and so they were manufactured through most of the nineteenth century. For forging applications

Trip-actuated helve hammers date back to at least the mid-seventeenth century in America. Despite the appearance of more sophisticated crank-actuated machines, these hammers were manufactured throughout the nineteenth century, and some remained in use throughout most of the twentieth century. (*Manning, Maxwell & Moore catalogue*)

which did not require adjustment of the hammer for changes in the thickness of the work, they served quite well. In fact, trip-actuated helve hammers could be found at work, through the early 1980s, in the heart of the high-tech "Silicon Valley," in San Jose, California. They were used in the shop of Elroy Christian to efficiently and profitably sharpen specialized agricultural components such as sugar beet spikes and bean harvester teeth.[3]

Another misleading term applied to some crank-actuated power hammers was the name "Justice Hammer." While it is true that the earliest hammers built in a vertical configuration (with a distinctive bow-spring linkage) were manufactured by the Philip S. Justice Company of Philadelphia, the actual inventor was Thomas Shaw. The Philip S. Justice Company continued to build hammers

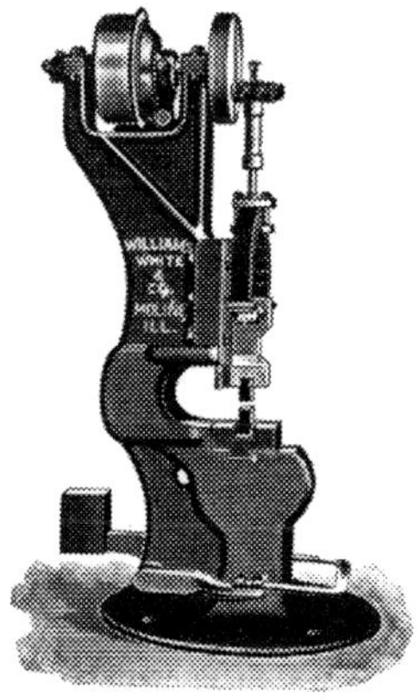

The Williams, White "Justice Hammer" was actually designed by Thomas Shaw. (*Williams, White & Company catalogue*)

of this type into the twentieth century, but the term "Justice Hammer" is perhaps most closely associated with the machines built by the Williams-White Company of Moline, Illinois. It had obtained a license to build the hammer in the early 1870's, and the Williams-White "Justice Hammer" was manufactured for many decades. By the mid-1950's, the origin of the name had been forgotten, and the company stated that "the Justice Hammer got its name from the inventor, a man of that name."[4]

The terms assigned to other types of open-die power forging hammers can be equally confusing. Designations such as "helve hammer," "tilt hammer" and "steam hammer," are imprecise, making comparisons of different machines problematic. Even the venerable Oxford English Dictionary offers little useful clarification of these terms. The term "steam hammer" is most often used to refer to the direct-acting steam-powered hammer which was developed in England and France in the 1840's. However, it can also be found in reference to earlier helve hammers driven by steam engines, as well as to the large steam-cylinder-actuated helve hammers (such as those built by the Cuyahoga Steam Furnace Company of Cleveland, Ohio, in the 1870's and 80's). Toward the end of the nineteenth century, many steam-powered hammers were converted to operate on compressed air. In fact, one of the largest of the giant "steam hammers," the 100-ton hammer built in Belgium by Cockrill, for the Terni Steel Works of Italy, was designed from the beginning to operate as a compressed-air hammer.[5] Among direct-acting steam/air-powered hammers there was also a variety of types. The earliest of these hammers were "single-acting," which is to say that the steam was used simply to raise the hammer, leaving the force of the blow to gravity. "Double-acting" steam hammers soon appeared, which used steam pressure above the piston to enhance the fall of the hammer. Double-acting steam hammers were thus far more powerful for a given hammer weight. Both single and double-acting hammers were built concurrently, and some machines could operate in either mode. The term "steam hammer" is thus by no means explicit.

Power forging hammers of all types are often described or specified on the basis of the weight of their falling parts. For example, the largest open-die hammer ever built (the single-acting steam-actuated hammer at Bethlehem Steel in Bethlehem, Pennsylvania) boasted a weight of 125 tons. This figure referred to the combined weight of the upper die and its key, the head or "tup," the rod, and the piston. The overall machine weighed more than 500 tons, stood 90 feet tall, and its anvil sat on a pyramidal mass of metal plates weighing an additional 1,800 tons![6] Similarly, among crank-actuated hammers, a Bradley Rubber Cushioned Helve Hammer would be specified by the combined weight of the upper die and key, the head, and its bolts (weighing altogether from 15 pounds to as much as 200 pounds). In reality, the weight of the falling parts is useful only when comparing machines of very similar construction. A 300-pound Bradley Upright Cushioned Strap Hammer is far less powerful than a 300-pound double-acting steam hammer. In both cases the force of the blow is augmented, but the added force of the steam is far greater than that of the Bradley hammer's mechanical springs.

The designation of hammers by weight has also been a source of confusion among historians unfamiliar with the industry. Victor S. Clark, in his classic, *History of Manufactures in the United States*, reports the installation of a 23-ton trip hammer at a railroad shop in Augusta, Maine, in 1869. According to Clark, the installation of this machine reflected a clinging to old usages and a questioning of the superiority of the direct-acting steam hammer. In actuality, it was the entire machine which weighed 23 tons, and the falling parts weighed closer to 1,000 pounds. Returning to Clark's source: since the hammer "...may be graduated to strike with a force anywhere from 2,000 pounds to one pound" it is likely that this machine was not a "trip" hammer at all, but a steam hammer, or a crank-actuated "atmospheric" hammer. In any event, the machine was state-of-the-art.[7]

In light of the confusion which has surrounded power hammer terminology in the past, it seemed wise to reconsider the fundamental, distinguishing mechanical characteristics of these various machines. Two primary points of distinction became evident: *actuation* (the means by which a reciprocating motion is communicated to the ram) and *geometry* (the linear or curved path traveled by the ram). The simplified classification which follows has proven extremely helpful in studying the open-die hammers built by more than one hundred and fifty American firms, as well as the hundreds of patents pertaining to these machines.

Classification of Open-die Power Forging Hammers

ACTUATION

1) TRIP
2) CRANK
   (For hammers with a guided-ram geometry, specify if actuating mechanism is vertically or non-vertically-configured. Specify linkage)
3) DIRECT
   A) cylinder and piston (steam/compressed air/ gasoline/gunpowder)
   B) foot-powered (not using a trip or crank actuation)
   C) other (electric, etc.)
   (If applicable, specify: counter-blow, self-contained, self-acting, single-acting or double-acting)

GEOMETRY

1) HELVE (arc)
   (Specify if the helve is of a compound, parallel-bar design)
2) GUIDED-RAM, -ROD, OR -CYLINDER
   (linear)
   (If applicable, specify: multiple cylinders, single-frame, or double-frame)

All open-die power forging hammers are actuated by one of three possible means: 1) a trip mechanism, 2) a crank (with some sort of spring cushioning or intervening spring linkage), or 3) "directly." The latter category includes primarily hammers which employ pistons in cylinders. They were most often powered by steam or compressed-air, but some used internal combustion (gasoline) and even gunpowder. Also belonging to this group (by virtue of the simplicity of their mechanism) are foot-powered "Olivers,"[8] as well as electric forging hammers (which existed, at least, on paper). In addition to the means of actuation, all open-die hammers possess a ram which moves either in an arc or in a straight line. The most common type of linear geometry results from a guided ram; however, some steam-actuated hammers employed a guided rod (when high clearance was required for jobs such as locomotive forgings). One unusual form of steam hammer, designed by the Englishman John Condie, used a guided cylinder (with the piston rod fixed). There are no instances of a hammer using both an arc geometry and a guided ram (though, practically speaking, the helve itself serves as a guide). There are, however, many examples of "helve-like" mechanisms used to communicate a reciprocating, linear motion from the crank to a guided-ram. In some cases, these non-vertically-configured, crank-actuated hammers clearly evolved from true helves (the Bradley Upright Strap Hammer, for example).

The Bradley Upright Strap Hammer used a helve-like beam to oscillate its guided ram. (*The Iron Age*)

However, with the linear motion of their rams, there is little practical difference between these machines and the vertically-configured hammers which simply employ a crank plate above the ram. While there are no guided-ram helve hammers, there is a clever variation of the helve which gives the smith parallel die faces throughout the hammer's stroke. The best-known hammer of this type was developed by William Blacker, of Stalybridge, England. It employed two parallel pivoting bars for the helve. The hammer

The Blacker Hammer used a distinctive parallel-bar helve which kept the upper die parallel to the anvil throughout the stroke. (*The Blacksmith & Wheelwright*)

ram thus rotated slightly through its arc, remaining parallel to the anvil at all times. Blacker's hammer was one of the very few foreign crank-actuated hammers to find a wide market in the United States. Since it was also manufactured in this country for a time, it will be considered here at greater length in the chapter devoted to crank-actuated helves.

* * *

It would be impossible to describe all of the varied tasks to which crank-actuated open-die power forging hammers were applied. However, in the twentieth century there were three specific jobs which were repeatedly mentioned in advertisements and discussions of these machines: 1) the sharpening of plowshares,

2) the welding of tires, and 3) the welding of axles. These common repair operations were laborious and often impossible to accomplish single-handedly. An increase in the demand for these specific services was undoubtedly responsible for the existence of many of the hammers which appeared in the first two decades of the century.

The plow is surely among the most important of tools. As the plow and kindred agricultural implements were improved, harvests increased and populations flourished. For a time, as plow technology evolved, the village blacksmith played an important part. He fabricated early wrought-iron-sheathed plows for local farmers; and, with the appearance of factory-produced steel plows, he assumed the burden of repair and maintenance. Demand for this branch of blacksmithing work grew steadily, and often it came in a rush just before planting time. Since "laying" a plowshare by hand can be laborious and time consuming, power hammers became increasingly attractive to blacksmiths in agricultural areas. Many of the small power hammers that appeared during the first decades of the twentieth century were built and marketed with plowshares prominently in mind. Today, plow technology no longer requires a ready forge and the skill of the plowsmith. Shares are hard-faced in the factory, and when worn beyond use, they are simply discarded and replaced.

The transition from the ancient wooden plow to one of iron was well underway by the beginning of the nineteenth century. This change did not occur easily. In Thomas Jefferson's time there was a widely-held belief among farmers that iron plows would poison the soil![9] Eventually, the advantages of a smoother, sharper, lighter and more durable plow won the day. Cast iron plows were readily mass produced, and the inherent brittleness of the material could be mitigated by a clever process which also provided a self-sharpening edge. So-called "chilled" iron plows were cast with metal inserts placed in the sand mold along the underside edge of the share. These "chills" caused the molten iron to solidify more quickly, yielding in their proximity a dense surface that was both harder and less susceptible to rust. When necessary, chilled iron plowshares could be reshaped by grinding, but they could not be heated and worked at the forge. Chilled iron plowshares were both cheap and durable (more resistant to wear, in fact, than some steels). They were extremely effective in the sandy, gritty soils of the Eastern states, but farmers in the Midwest were less than satisfied.

As America expanded westward the great plains were brought under cultivation. Farmers quickly discovered that their trusted chilled plows would not "scour" properly in the smooth soil of the prairie. In other words, the soil would stick to the surface of the mouldboard (the portion of the plow extending above the share) creating drag and inhibiting the formation of an even, well-pulverized furrow. The solution to this problem made John Deere rich. He substituted steel plate for cast iron, thereby creating a plow which would scour in prairie soil. The first billet of "cast plow steel" was rolled for Deere in Pittsburgh in 1842. By 1858, at the time when crank-actuated power forging hammers first appeared, the year's production of John Deere plows was 13,000.[10] Plows like John Deere's celebrated Moline Plow used steel shares as well. They were lighter and more efficient than cast iron plows, but they were also more expensive. Since they lacked the self-sharpening characteristic of the chilled plows, it became common practice for the plowman to carry a large file or rasp into the field.[11] According to *File Filosophy* (espoused by the Nicholson File Company of Providence, Rhode Island), the contribution of plowshares, seed

The celebrated Moline Plow, built by John Deere, used a steel share as well as a steel mouldboard. This example has a disc coulter to assist in breaking through the soil. (*Farm Implements and Farm Machinery*)

A small Kerrihard Power Hammer is used here to draw out the edge of a worn plowshare.

drills, cultivator shovels, harrow spring teeth and discs is assured "when 'whipped' into faster working action by the unsparing use of the faithful file."[12] The natural abrasion of the soil as well as frequent dressing of the edge removed material and gradually weakened the crucial point and edge of the share. If a plowshare was merely dull, the blacksmith might carefully hammer the edge to sharpness; but if it was well worn, it became necessary to "lay" a new edge and point onto the share by forge welding a steel addition. Even after the appearance of relatively cheap steel, the cost of restoring a steel share at the forge was still less than that of replacement. Additionally, the village blacksmith served as a plow technician, adjusting and maintaining the implement to suit each farmer's particular needs.

Despite its apparent simplicity, a plowshare is actually quite sophisticated. Not only does it cut the soil, but the geometry of the share determines the degree of "suction" which holds the plow bottom at the proper depth. Many styles of shares were devised for specific conditions and various types of steel were employed. The most sophisticated steel shares used a laminate which consisted of hardened layers surrounding a soft steel core. This design provided a self-sharpening edge much like that of the chilled cast iron. One such laminated steel was produced by the Crucible Steel Company of America. Its "La Belle Soft Center Plow Steel" (named for Crucible's La Belle Works in Pittsburgh) was available to plow makers in "Bars, Slabs and Shapes, cut to pattern, for use in the manufacture of Mouldboards, Shares, Landsides, and Cultivator Shovels..." This steel was manufactured using open-die steam-actuated hammers; and was available with either hammered, hammer finished, or rolled finish. A similar technology could be observed in Crucible's "Five-ply Tool-proof Safe and Vault Steel" and "Seven-ply Screw Steel," both of which employed hardened layers of chrome alloy steel.[13]

Laying a share demanded the highest degree of control in forge welding, as well as a high standard of craftsmanship in finishing the smooth surfaces that were required to scour cleanly in the soil. Thus, it was easy for an inexperienced blacksmith to ruin a plowshare. The shape of the share was crucial, as was the proper heat treatment. Even if everything was done correctly, the American farmer was sometimes difficult to please. In *Heartland Blacksmiths*, Leslie Ostendorf, an agricultural blacksmith from Addieville, Illinois, recounted his first experience sharpening plowshares. He had watched the owner of the shop sharpen shares, and did his best to emulate the process. When the farmer returned the shares, claiming that they would not stay in the ground, the owner looked them over. He found that the only thing amiss was that the farmer was suspicious of a new smith. When the shares were reblackened in the fire, the farmer was satisfied with the result.[14]

At the beginning of the twentieth century, America was still moving largely upon wheels of wood. The durability of wooden wheels was dependent upon a continuous band of iron which both held the wheel together and resisted the abrasion of the road. Under the less-than-perfect road conditions of the day, these iron tires wore out quickly. Furthermore, if there was any stretching of the metal or contracting of the wood, it became necessary to adjust the rim. Tires could be made slightly smaller using a device known as a "tire shrinker." The shrinker gripped the band and effectively upset it slightly using a squeezing action. If the tire was to be

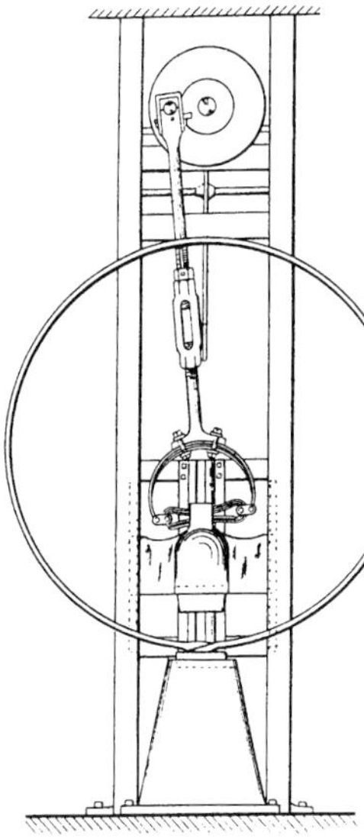

This drawing, which accompanied T. H. Griffitts' 1904 patent (number 773,735 which was granted on November 1), illustrates the use of a power hammer to weld a wagon tire. Griffits was from San Francisco, California, and it is not known if the design was ever produced commercially.

replaced, it was necessary to weld a new band of the proper diameter. Small tires, such as those of light buggies, were most easily welded by hand. Wider, thicker tires were another matter. Heavy-duty farm wagons often used tires four inches wide by five-eighths or even three-quarters of an inch in thickness. A 48-inch diameter tire of these dimensions was extremely cumbersome, and rapid heavy blows were needed to accomplish the weld. Power hammers were well suited to such circumstances, eliminating the need for multiple human strikers. Some hammer builders provided special dies for tire welding; and one manufacturer, Kane and Roach of Syracuse, New York, incorporated a rest to support the tire as it was being welded. Dedicated tire-welding hammers were built by Williams, White and Company of Moline, Illinois. Its 55-pound hammer was recommended for welding tires up to four inches by five-eighths of an inch, and a 90-pound model was offered for tires up to six inches in width. With special dies these hammers could

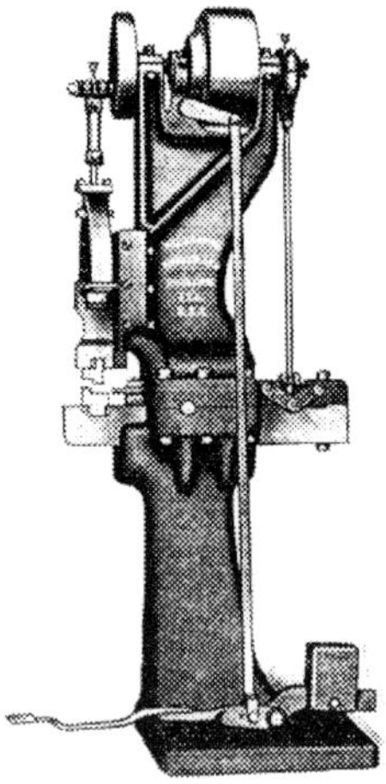

Williams, White and Company of Moline, Illinois, built a line of dedicated tire-welding hammers. This 55-pound model included a horizontal "side stroke" feature for edging. (*Williams, White & Company catalogue*)

also weld rims for steel-spoked wheels. The use of such hammers by wheel manufacturers ended soon after the turn of the century with the application of electrical resistance welding; but in repair shops, forge welding with a power hammer persisted for several decades as the most practical method for welding tires.

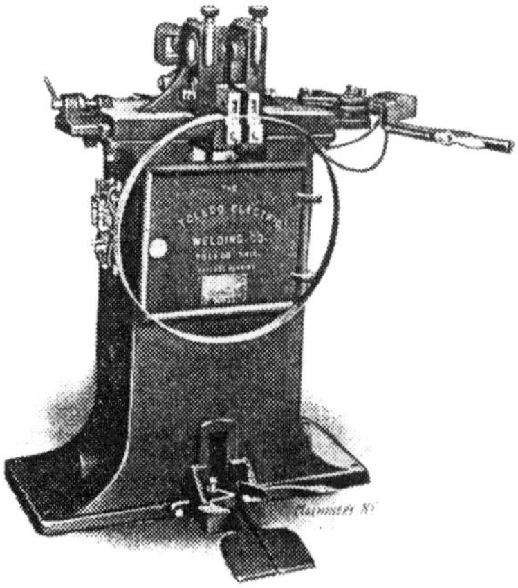

Electrical resistance welders, such as this one built by the Toledo Electric Welding Company of Toledo, Ohio, replaced hammers for tire welding in the factory. (*Machinery*)

In addition to welding tires, another aspect of carriage work which often confronted the general blacksmith was the replacement of worn axles. Most vehicle axles in the days before the high-speed automobile were simple. Instead of roller- or ball-bearings (which can be readily replaced without replacing the axle itself) a typical farm wagon or buggy relied upon a well-greased steel spindle running in a cast-iron box. Wagons with simple wooden axles employed "thimble skeins" (generally of cast iron or forged wrought iron) which fit over the tapered end of the wood. The turn-of-the-century blacksmith might well stock replacement skeins for his customers. However, many vehicles were built upon solid steel axles, with the tapered spindle sections machined at the ends of a single bar. When such a spindle became worn or deeply scored it was necessary to cut the old spindle off and forge-weld a new spindle stub in its place. Larger axles of this type could pose a challenge to the rural smith. Not only did the task require welding skill and strength, but finesse was needed to assure that the resulting axle was of the proper length. Therefore, it was not unusual for certain shops to specialize in this job. The Cray Brothers of Cleveland, Ohio, sold stub axles in their 1909 catalogue of carriage and wagon tools and materials, but they also offered to perform the repair for the blacksmith. A set of stubs for repairing a one and a half inch axle cost the smith $3.21, and the additional charge to weld and set the axle was $1.50. If this was a frequent operation for

AXLES—Half Patent Double Collar

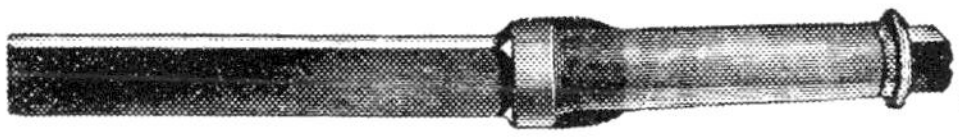

MADE OF BEST REFINED STEEL

Spindles Turned Perfectly True and Polished

No. 12. SHORT BED OR "STUBS"

Beds are about 8½ inches long, from collar. Suitable for welding to old axles.

	Per Set		Per Set
¾ x 6 and 6½	1.12	1⅛ x 6½ and 7	$1.48
⅞ x 6 and 6½	1.15	1¼ x 6½, 7 and 7½	1.80
15/16 x 6 and 6½	1.16	1⅜ x 7 and 7½	2.44
1 x 6 and 6½	1.28	1½ x 7½ and 8	3.21
1 1/16 x 6½	1.35		

The Cray Brothers of Cleveland Ohio sold axle stubs to blacksmiths, but also offered to weld the axles in their own shops. (*Cray Brothers catalogue*)

a smith, it is easy to see how a power hammer which cost between $60 and $100 might well pay for itself in less than a year. Even at the beginning of the century, simple axles of this type were on the way out. The same Cray Brother's catalogue of 1909 also includes sets of Timken roller bearings which could be retrofitted into old wooden hubs in lieu of the cast iron boxes.[15]

* * *

While most blacksmiths today are familiar with a variety of power forging hammers, most will undoubtedly be surprised by the sheer number of different hammers that were built in America. The profusion of hammers which appeared in the early twentieth century is particularly remarkable. Small crank-actuated hammers (in the 25- to 50-pound range) were given colorful names, such as the Boss, the Common Sense, the Little Giant, and the Comet; and marketing of these machines assumed much the same character as that of more conventional consumer goods. In fact, household appliances such as washing machines from this period were often given these same names! This widespread development of crank-actuated power forging hammers was due to a number of factors.

Generally speaking, American manufacturers, including blacksmiths, were less conservative than their European counterparts. When Henry Bierhart of the Bradley Company visited Germany on a sales trip in 1877, he was struck by the almost exclusive use of trip-actuated hammers in the forges of edged-tool makers there.[16] American manufacturers were willing to experiment with new machinery, often placing new hammers side-by-side with older machines. A glimpse into the forge shop at the Green River Works of the John Russell Cutlery Company (in Turner Falls, Massachusetts), ca. 1890, reveals a selection of state-of-the-art crank-actuated hammers in a row with trip-actuated helves.[17] The profusion of new power hammers that appeared on the market in the early twentieth century reflected the extent to which American blacksmiths were willing to move beyond a reliance on hand labor. Of course, there were many practical economic reasons for adding a

The forge at the Green River Works of the John Russell Cutlery Company, ca. 1890. Beyond the row of trip-actuated helves are at least three different models of crank-actuated hammers built by the Beaudry and Bradley companies. (Photograph courtesy of the Russell-Harrington Cutlery Company)

new gas engine and power hammer to one's shop; but fundamentally, the American blacksmith was open to change. Periodicals, such as *The Blacksmith and Wheelwright* encouraged this by offering prizes to their readers for contributions on such topics as the value of power in the shop and the use of "trip" hammers.[18]

Most of the hammers built during the nineteenth century came from Eastern manufacturing centers, but the vast majority of the smaller, cheaper power hammers which appeared in the early twentieth century were built in the Midwest. The first crank-actuated power forging hammers were developed at a time when steel was still a costly and exotic material. That soon changed with the application of first the Bessemer process and then the open-hearth process of steelmaking. Fuel costs were reduced by the shift from increasingly scarce charcoal to abundant coal. Exploitation of the iron reserves of the Lake Superior region of the United States was in its infancy when the first crank-actuated hammers were built; however, the successive openings of the Marquette, Menominee and Mesabi ranges were to have an enormous influence on American industry and hammer building. The proximity of such vast reserves of iron (for cast iron and steelmaking) caused heavy industry to move westward. Favorable raw materials prices, and cheap transportation by rail, made it possible for small foundries to flourish throughout the Midwest in previously isolated agricultural areas. At the same point in time when many smiths were first adding a source of power (generally a gasoline engine) to their shops, foundrymen were eager to develop new products to expand their businesses. Many, in fact, built gas engines; others made mills for grain and products such as the ubiquitous cast iron water pump. Few of the companies which poured castings for power hammers, and assembled these machines in their shops, were devoted exclusively to this single product. The extent to which the American foundry business had become decentralized (and practically portable) is evident in the fact that one foundryman-hammer-builder, Arthur Alfred Koch, was able to move his business from Albert Lea, Minnesota, to Waterloo, Iowa, to Austin, Minnesota, and back to Albert Lea in a period of only four years! This record also reflects the fact that while the foundry business and the specialized hammer market was robust, it could also be extremely challenging.

* * *

In the chapters which follow, each of the three principal types of crank-actuated hammers is considered separately. The stage will first be set, though, with an account of the earliest hammers to employ a crank actuation: the so-called "atmospheric" hammers.

A human striker provides the brawn for a swaging operation. (*International Library of Technology*)

Bennet Hotchkiss' Atmospheric Hammer was crank-actuated and used an air-cylinder spring linkage. (*The Scientific American*)

CHAPTER 2

"Troublesome Implements" and Ingenious Solutions

> "These hammers are troublesome implements. For shingling blooms, they can be replaced by squeezers; but they are required for drawing bar iron... An ingenious inventor has a fair field for improving the present machinery."[1]

So wrote Frederick Overman in 1849, describing the trip-actuated helve hammers then in use at most American forges. This was not the first call for improvement in forging machinery. In 1810, Tench Cox described blacksmithing in America as,

> "a branch of manufacture of great amount and utility...performed in every section of the United States, however agricultural, and however peopled. *The improvement of the system and tools of these workmen, and particularly the introduction of labor-saving machinery into their shops, would be a very substantial benefit to them and to the country*." (Cox's own italics).[2]

During the first half of the nineteenth century there were, in fact, modest improvements. The importance of iron and steel to the young nation increased steadily, and the demand for forgings swelled. Forgings for mass-produced agricultural implements, steam engines, locomotives, arms, textile machinery, etc. challenged the state-of-the-art, and prompted a variety of important developments. Most notable of these was the development of closed-die drop forging. Such a drop hammer with dies was used by John Hall in 1827 at Harper's Ferry for forging rifle parts. The head, or ram, was raised by a chain and released at the desired height by a trip.[3] By the middle of the century this technology was developing rapidly, moving beyond the esoteric realm of firearms manufacture. In 1851, Milo Peck of New Haven, Connecticut, patented a crank lifter for drop presses and hammers.[4] Friction rolls soon appeared, raising drop hammer rams first by leather belts and then by boards of wood. Shortly after the Civil War, the firm of Billings and Spenser began to manufacture its renowned board drops.

Some technological changes, such as the increasing use of cast iron in the construction of textile machinery and machine tools, led directly to similar developments in forge hammers. At least one important manufacturer of textile machinery, the Lowell Machine Shop, of Lowell, Massachusetts, also built small, iron-framed trip hammers. Improved trip-hammers found immediate application in forging spindles by the thousands for use in the mills. In fact, some early styles of cast-iron-framed trip hammers were known by the names of the textile-producing centers in which they were manufactured: the "Lawrence," "Holyoke" and "Chicopee."[5] Robust, cast iron frames improved the durability of the trip-hammers which were used in simple, repetitive, manufacturing roles, but the fundamental limitations of the design remained. The dies were parallel at only one position, variations in stock thickness required painstaking adjustment of the dies (through shimming), and the intensity of the blow was generally limited to the force of gravity. The force of the blow was thus diminished as the thickness of the work increased (often in precise opposition to the smith's needs); and, additionally, the trip actuation was extremely hard on the hammer itself and its drive. They were indeed "troublesome implements."

The direct-acting steam hammer, which had been developed in England during the 1840s, was also troublesome in the eyes of American forgemasters. Several of the steam hammers built by James Nasmyth at Patricroft in England were soon at work in America. Also, the rights to construct hammers under Nasmyth's patent were almost immediately acquired by Merrick and Towne, of Philadelphia.[6] While these hammers quickly proved themselves indispensible for producing the largest forgings, such as steam engine shafts, smaller hammers built on this principle were simply too slow for drawing-out iron and for producing small-scale work. Frederick Overman considered the Merrick and Towne hammer "a fine implement," but observed that "they make but 80 or 100 strokes per minute; while it is necessary that a hammer suitable for drawing iron should make at least 150 strokes; and smaller hammers from 150 to 300, and even 400 strokes per minute."[7] In addition to this technical limitation, the fee to build a hammer under Nasmyth's patent was almost prohibitively high (360 British pounds, or $1,800).[8]

In a search for alternatives to the patent-protected and slow direct-acting steam hammer, American inventors began to explore other ways of using a cylinder and piston to operate a forge hammer. In the steam-powered hammer, steam pressure was used to raise the ram, propel it downward (if the hammer was "double-acting"), and also to precisely regulate the force of the blow. Additionally, the cushion of air in the cylinder served to isolate the machine somewhat from the shock of repeated blows by the ram. Bernard Hughes, of Rochester, New York, rose to the challenge and patented a trip-actuated hammer which accomplished most of these things without requiring steam pressure. In Hughes' design (patent number 10,923 granted May 16, 1854), the rod which bore the hammer ram also held a piston plate which moved in a cylinder

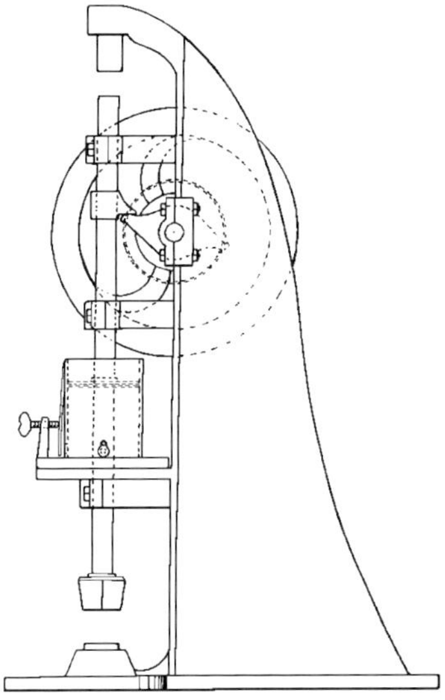

Hughes' patent of 1854 described a trip-actuated atmospheric hammer.

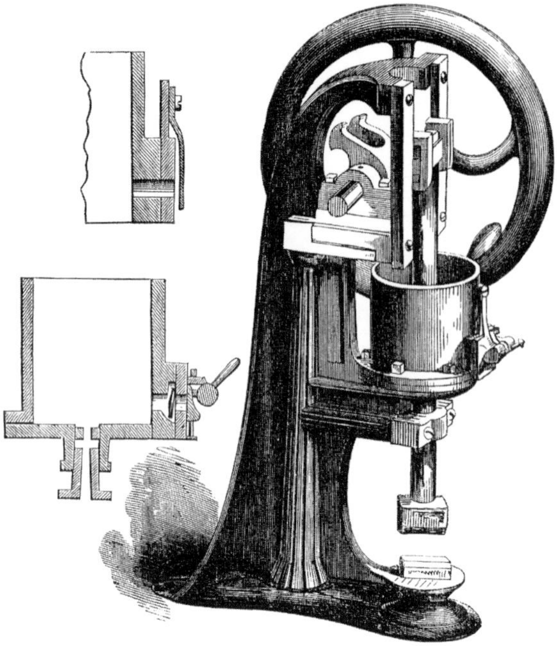

Hughes' Atmospheric Trip Hammer of 1854. (*The Scientific American*)

controlled by valves. The hammer ram was raised and released by a simple trip mechanism, but the downward force of the ram could be increased by creating a vacuum under the piston. It was also possible to regulate the force of the blow, diminishing it by trapping air under the piston as the ram fell. As with the steam-powered hammer, this arrangement provided some isolation from the deleterious effect of repeated impact. Hughes' invention was deemed important enough to warrant front page coverage by *The Scientific American*, appearing in its issue of November 11, 1854. At that time, hammers of Hughes' design had been in operation in Rochester for more than six months. Considered fully tested, the Rochester Iron Works claimed to be turning out six hammers per week. Hughes' hammers were built in four different sizes "so as to adapt them for heavy and light work - for forging iron and steel, and for hammering brass, tin, and copper, by tin- and copper-smiths."[9]

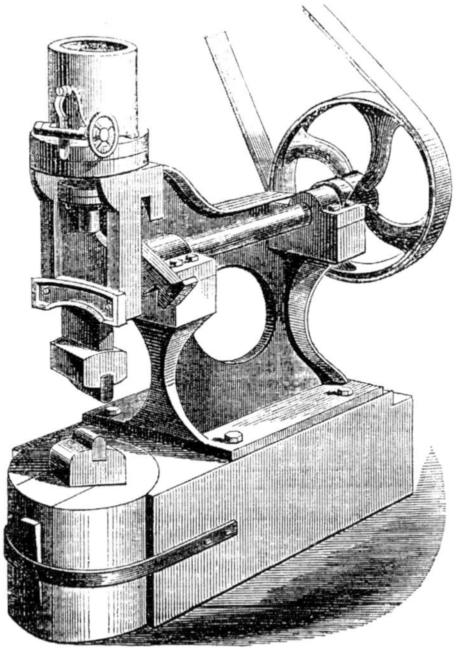

Hughes' Atmospheric Trip Hammer, 1859. (*The Engineer*)

Hughes' Atmospheric Hammer quickly became a success. In 1858, at the Fair of the American Institute, one of Hughes' hammers was observed deftly cracking filberts after having just delivered a series of the heaviest blows - a spectacle reminiscent of steam hammer demonstrations in England.[10] In November of 1859, *The Scientific American* again profiled the machine, this time as an "Improved Atmospheric Hammer." Changes had been made in the operation of the valves, the location of the trip-actuating cams, and india-rubber cushions had been placed under the journal boxes which held the drive shaft.[11] This early engineering application of vulcanized rubber was apparently necessitated by an ongoing problem with the shock caused by the trip mechanism. With these refinements, Hughes' hammer began to attract attention overseas. The hammer, which was now being built at the vast Novelty Works in New York City, was profiled in *The Engineer* (London).[12] Though

Hughes had patented his invention abroad, several variations on the theme soon appeared. A substantial atmospheric hammer built by Waterhouse in Sheffield, England, was described in *Dr. Dingler's Polytechnischem Journal* (Berlin) in 1859.[13] The next year, *The Engineer* illustrated and described another English atmospheric hammer patented by George Dawes and Charles J. Carr of Hoyland, Yorkshire.[14] Atmospheric hammers flourished in England; and remarkably, after a quarter-century had passed, an improved version was re-introduced to the United States in the form of the Hackney Power Hammer. This hammer was built for a brief time in Cleveland, Ohio, and will be described in the chapter dealing with vertically-configured hammers. Atmospheric hammers were also influential in the development of self-contained air-actuated hammers, but that is another story.

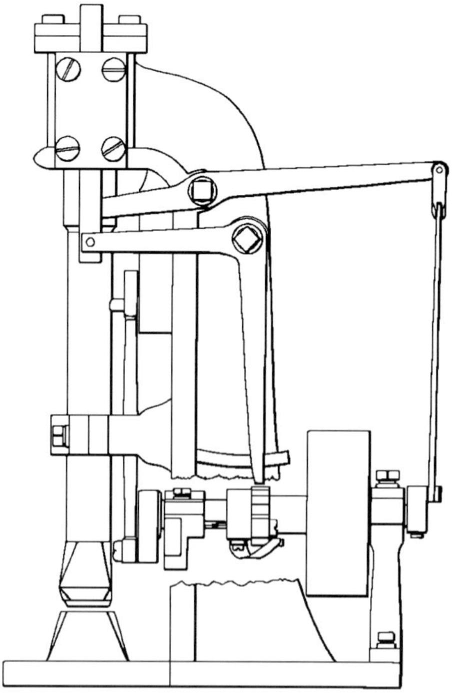

Milo Peck's 1855 patent described the first crank-actuated atmospheric power forging hammer.

Peck's Lifter ("latest improved," 1884), was first patented in 1851. (*City Directory*, courtesy of the Whitney Library, New Haven)

The immediate influence of Hughes' Atmospheric Trip Hammer in America was profound. Eight months after the initial description of the invention in *The Scientific American*, Milo Peck of New Haven patented his own version of an atmospheric power hammer. Peck's invention (granted patent number 13,179 on July 3, 1855) used a sophisticated automatic valve arrangement to control the air cylinder. Unlike Hughes' design, which employed a vacuum or air-cushion created *below* the piston to control the hammer's stroke, Peck's hammer used a valve controlled by two sliding plates to regulate the flow of air *above* the piston. Air compressed on the upward stroke propelled the ram downward, and a vacuum created above the piston controlled the extent to which the ram could descend. Since Peck had prior experience and commercial success with a crank mechanism (as applied to drops),

Milo Peck & Co. advertisement from 1873. (*Benham's New Haven City Directory*, courtesy of the Whitney Library, New Haven)

he quite naturally used the same technique to actuate his rapid-blow, open-die hammer. Peck's 1855 patent marks the first description of a crank-actuated open-die power forging hammer. While the design was never placed into production, it was undoubtedly influential.

The drive shaft of Peck's hammer, which was belt-driven from a line shaft, extended to the rear of the machine. The actuating crank was located just behind the work area,* and a valve (worked by two sliding plates) regulated the flow of air into the cylinder. One of these plates, controlled by a hand lever, set the valve opening, while the other plate opened and closed the valve cyclically (driven by a cam on the rear of the drive shaft). The actuating crank was attached to the hammer rod by a connecting link without any sort of flexible linkage to absorb shock. This was quite unlike Peck's Patented Lifter, which used a leather strap to link the crank pin and the drop. In addition to its crank motion, the lifter was designed in such a way that the lifting action increased slowly and then decreased prior to release (minimizing wear and tear on the lifter itself).

New Haven had risen to become an important center of American manufacture in the decades prior to the Civil War. Its earliest industries focused upon carriage building; but by 1855, firearms and hardware were prominent. An English committee studying machinery and manufacturing methods in the United States visited the Whitney Arms Company there, as well as several highly-mechanized factories producing locks, clocks, and india-rubber shoes.[15] Both the Winchester Repeating Arms Company and Sargeant and Company (hardware) became established in New Haven during the 1850's. Amidst this activity, machinist and inventor Milo Peck prospered. He had formed his own company in

---

* Technically, this would make the machine a "non-vertically configured" hammer in the author's scheme; but since most machines of that type used a pivoting beam, it is just as well that it is considered here, separately. It is interesting to note that, in his patent, Peck mentions the possibility of positioning the lifting device above the air cylinder - a vertical configuration.

1850, and quickly rose to prominence with the development of his crank lifter. It was applied to both sheet metal and forging drops "for the manufacture of hardware, cutlery, gun parts, spoons and forks, silver tableware, lamps and lanterns, tinware, steel plows, horseshoe nails, jewelry, tools, toys, and thousands of other articles."[16] Milo Peck and Company was succeeded by Beecher and Peck (Henry M. Beecher & George W. Peck); which, in turn, became the Miner and Peck Manufacturing Company. The penultimate Peck Manufacturing Company disappeared from the scene in the early 1920's.

From an early point in the 1850's, Milo Peck manufactured both his lifters and the drop presses and hammers that used them. An advertisement from 1857 notes that Peck was also making oval chucks (for use with turning and spinning lathes); but, significantly, it fails to mention forging hammers.[17] Apparently, the development of Peck's atmospheric hammer had not proceeded smoothly. The nature of the difficulty was described in Peck's second hammer patent (number 27,925 granted April 17, 1860):

> "A practical trial of the machine constructed on the principle of the one referred to [in patent number 13,179], demonstrated the existence of some serious difficulties in the way of its success, the most important of which was the injury which the various parts of the mechanism sustained from the shock occasioned by the strain on the connecting bar in overcoming the inertia of the hammer - the danger of injury from this cause being, in practice, increased in proportion to the height to which the hammer was raised above the face of the anvil by the object to be hammered, from the fact that the connecting rod moves at an accelerated speed after the crank has passed the center until it reaches the half stroke, and the usual thickness of the article to be worked being less than one half of the stroke of the piston."[18]

Peck's solution to this difficulty was to interpose a cushion of india rubber between the hammer rod and the cross head block (which connected the ram to the pitman). Also, since he felt that it was important for both the hammer mechanism and the anvil to share a single bed-plate (to preserve registration of the dies used for swaging

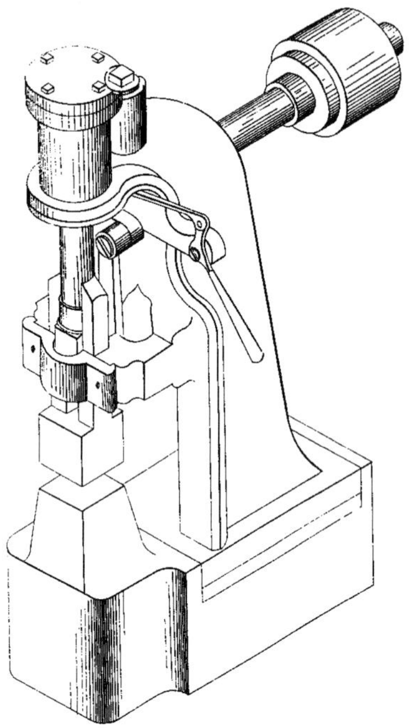

Milo Peck's patent of 1860 described improvements which included the addition of a rubber cushion between the piston rod and the ram.

work "into peculiar forms"), Peck also described a means of isolating the hammer from the bed-plate with thick planks of wood "or some equally good non-conductor." While Peck's improved design was doubtless superior to the initial concept, another New Haven inventor had already patented what was to become the first crank-actuated open-die power forging hammer to be successfully manufactured.

* * *

Bennet Hotchkiss envisioned a different sort of atmospheric hammer. Both Hughes and Peck had steam hammers firmly in mind when they incorporated fixed air-cylinders into their machines. These air-cylinders were precisely regulated by valves which enabled the hammer to strike with the greatest possible control. Hotchkiss, on the other hand, placed the air-cylinder in motion. It became primarily a spring, using a simple valve which functioned automatically as the air-cylinder moved up and down. The Hotchkiss Atmospheric Hammer, in its simplicity, was far more practical than its more sophisticated predecessors. Milo Peck had also considered using an air-cylinder linkage (in lieu of india rubber); but, even after the appearance of Hotchkiss' design, he expressed skepticism about its practicality. In his patent of 1860 he wrote:

> "I have experimented with various elastic bodies, but have found none which would combine the advantages of economy and utility so highly as india rubber. An air cushion produced by a piston on the cross head block working in a socket on the hammer shank would answer a good purpose but would be objectionable on the score of increased expense of construction, and liability to derangement, without any compensating advantage."[19]

As president of the New Haven Mechanics' Club, Peck was an important figure in the New Haven engineering community. He doubtless knew Bennet Hotchkiss personally, and was well-acquainted with the development of the Hotchkiss hammer. While there were indeed significant technical difficulties to be overcome in the use of an air-cylinder spring, Peck was quite wrong about the compensating advantages. The use of an air-cylinder spring (between the hammer ram and its actuating crank) may not have been the best solution in the design of such a hammer, but it remained in production for the rest of the nineteenth century.

Bennet Benedict Hotchkiss was born on August 15, 1829. His family, like Peck's, was well-established in New Haven, having arrived there seven generations before (prior to 1641). There is, in fact, a likelihood that the two inventors were distantly related,[20] though no details are known about their personal relationship. In 1850, when Peck was starting out in his own business, Hotchkiss was described as a "coach trimmer." He was listed as a "pattern-maker" the next year, and for several years he was employed by the New Hampshire Pattern and Machine Company. Hotchkiss subsequently entered a partnership with Francis S. Collins, with whom he shared the rights to his first hammer patent (1859). By the next year, Hotchkiss felt prosperous enough to marry. He remained in New Haven through the early 1880's, but little is known of his personal or professional activities beyond his hammer patents.[21]

Hotchkiss' Atmospheric Hammer, the first commercially-successful crank-actuated machine of its kind, was laid-out much like the hammer described in Peck's 1855 patent: the crank was located behind and slightly above the work area. The drive pulley was engaged by tightening a running leather belt by means of an idler pulley. The crank arm was connected to the rear of the air cylinder which traveled up and down in the same guides used to hold the hammer ram. The piston rod was connected to the ram. There were no adjustable valves for the air-cylinder, merely a small hole at its middle. This hole allowed air to enter each side alternately as the air trapped in the opposite end was compressed. The design was extremely simple and clever: energy was stored by the compression of air in the cylinder, propelling the ram upward after each blow, and then downward with added force. The same action could be achieved with metal springs, but Hotchkiss failed to anticipate the various spring configurations which were possible.

Hotchkiss' first hammer patent, number 24,428 was granted on June 14, 1859. In a re-issue of that patent (re-issue number 2,389 granted November 6, 1866) Hotchkiss sought to expand his claims to include a steel coil spring in lieu of the air cylinder. However, by that time, Thomas Shaw of Philadelphia had already developed and patented the use of an elliptical, or bow-shaped, steel spring linkage. When Thomas Shaw re-issued his own patent of 1866 (re-issue number 7,475 granted January 23, 1877) he noted the inadequacy of the spiral (coil) spring as being, "very liable to early fracture by the severe torsional strains of a vibrating hammer..."[22] A successful coil-spring linkage was not to appear until 1890, in the design of Phillipe Denery Dupont.

Bennet Hotchkiss' second patent, number 39,924 was granted on September 15, 1863. It covered numerous practical refinements to his initial design, including the addition of two poppet-valves to the air cylinder. These valves allowed the quantity

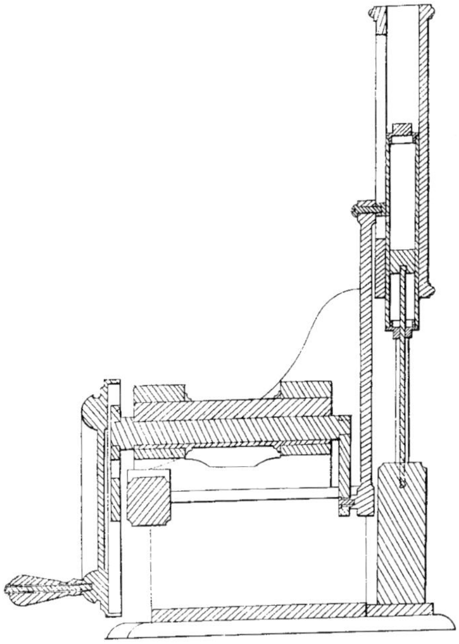

Bennet Hotchkiss' important patent of 1859 was illustrated with a hand-cranked model.

of air in either side the cylinder to be adjusted, so that the stroke of the hammer could be changed without altering the velocity of the hammer (a subtle limitation of the initial design). The poppet-valves were controlled by adjustable inclined planes, across which they travelled. By the time of Hotchkiss' second patent, the hammer was already in limited production. *The Engineer* (London) featured the Hotchkiss Atmospheric Trip Hammers in an article published in May of 1863.[23] An actual Hotchkiss hammer was illustrated by *The Scientific American* in October of 1864, and it was noted that one was being used "on gun work" by the Starr Arms Company of Yonkers, New York.[24] In September of 1865 it was claimed that more than forty of these hammers were in use in different parts of the country.[25]

As Milo Peck had suggested in 1860, the practical development of Hotchkiss' air-cylinder spring linkage was to be a struggle. When *The Scientific American* introduced Thomas Shaw's

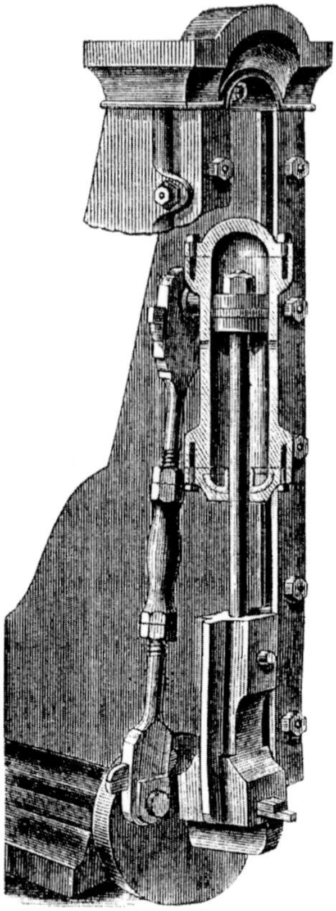

This detail of the Hotchkiss Atmospheric Hammer clearly shows the crank-actuation, adjustable pitman, and air-cylinder spring. (*The Scientific American*)

hammer, it was pointed out that atmospheric hammers were also in use at the time, " but each and all of these have peculiar difficulties inherent in them which cannot be surmounted."[26] The main problem with Hotchkiss' initial design was succinctly described in Edward A. Raymond's 1865 improvement to the Hotchkiss hammer (patent number 51,267 assigned to himself and the firm of Charles Merrill & Sons, builders of the Hotchkiss hammer):

> "In hammers of this character heretofore constructed the piston rod has been packed with an ordinary elastic packing, which produced so much friction that the air in the cylinder becomes almost inoperative in giving an elastic movement to the hammer,

because the friction is sufficient, or nearly so, to lift the hammer, and this character of packing quickly becomes leaky and useless in consequence of the heat and particles of dirt that adhere to the piston rod. The piston itself has been formed with cup-shaped leathers; but the cup leathers, if the movement is such that they touch either end of the cylinder, are forced out of shape and the hammer rendered inoperative. The hammer of the character referred to must be durable and not subject to derangement, because they are generally in the hands of blacksmiths, who would not understand their repair in case of their becoming out of order."[27]

While *The Scientific American* did not necessarily share Raymond's sorry lack of faith in the mechanical abilities of the American blacksmith, atmospheric hammers were judged, like steam hammers, as being troublesome and requiring frequent maintenance. But while steam hammers were successful in their niche, pneumatic hammers were ill-fated because, it was reasoned, "air is a more subtle fluid than steam."[28] Ultimately, hammers actuated directly by that "subtle fluid" were to dominate the field.

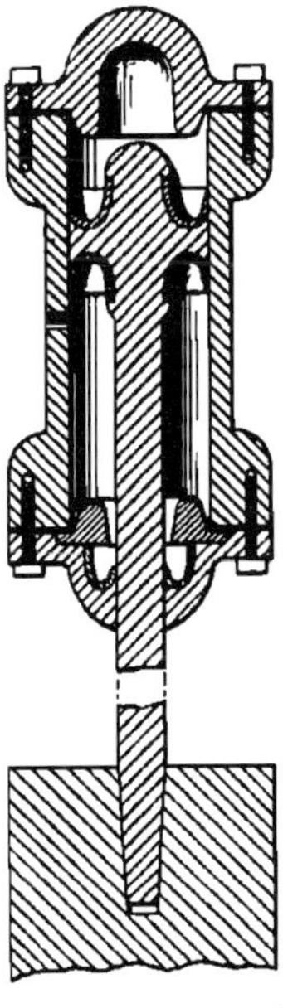

Edward Raymond's patent of 1865 describes an improvement to the leather packings used in the air-cylinder spring of the Hotchkiss hammer.

Bennet Hotchkiss was granted at least two further patents for forging hammers. Patent number 41,775 (granted March 1, 1864) pertained to a drop hammer design. Patent number 42,581 (granted May 3, 1864) explored an interesting approach to increasing the force of the blow in an open-die atmospheric hammer. By arresting the motion of the hammer ram at its highest point (or some other point chosen to provide the desired force), the full effect of the compression of air in the cylinder could be exploited. To accomplish this, the cylinder became an extension of the ram and the crank was connected to the piston. A trip lever automatically released the cylinder when the piston reached its lowest point. A foot-controlled plate moved against a poppet-valve near the bottom of the cylinder to regulate the amount of air, and thus the intensity of the blow. This was a clever approach, but it is not known if it was ever incorporated into actual, working hammers.

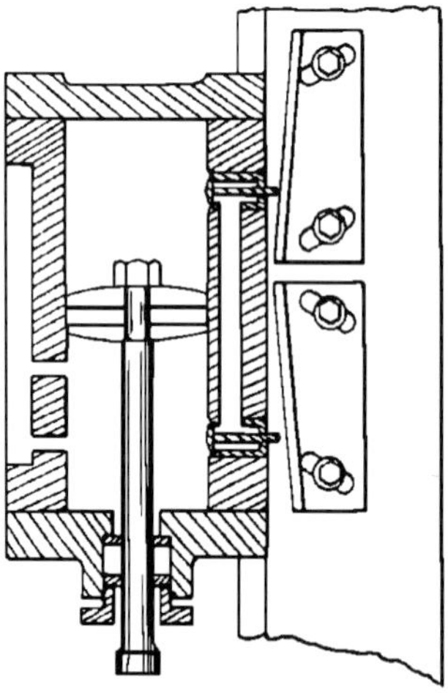

Bennet Hotchkiss' patent of 1863 described a far more sophisticated means of regulating air flow. Two additional poppet valves allowed the stroke to be varied without changing the hammer's speed.

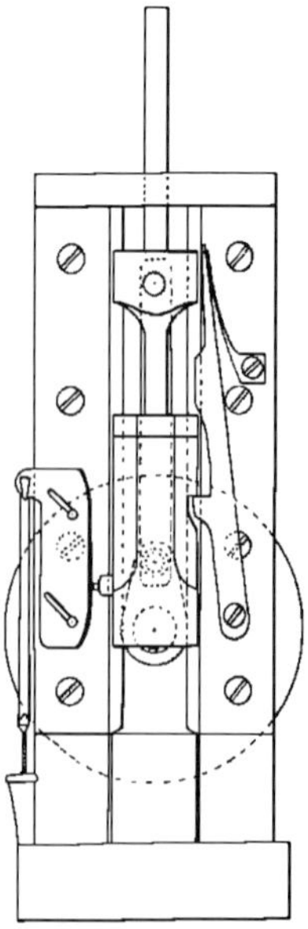

Bennet Hotchkiss' patent of 1864 described a means to delay the descent of the ram to coincide with the downward movement of the crank (maximizing the effect of the cylinder's compression).

The firm of Charles Merrill and Sons of New York City was the sole manufacturer of the Hotchkiss hammer, and it was produced for many decades. The Hotchkiss hammer, like many of the open-die power forging hammers which were to follow, benefited greatly from a period of heightened war production. In 1866, Charles Merrill and Sons claimed to have sold 80 hammers over the previous twenty months, "to large and small concerns." At that time, the hammers were offered in sizes of 10-, 30-, 40-, 50- and 60-pounds. It was said that "the 40-pound hammer will draw a three-inch bar three feet at one heat. It can be used in any building without injuring the foundation or walls, as it runs light and without noise."[29] By 1892, the Merrill "Upright Power Trip Hammer," as it came to be called, was offered in four sizes (25-, 50-, 80- and 100-pounds). An advertisement in *The Journal of Railway Appliances and Railway Price Current*, pointed out that "occasional attention to packings insures a perfect spring." The advertisement went on to suggest that replacement of the leather gland packing (used "in the same way a steam engine rod is packed") could be accomplished "by any person of ordinary intelligence."[30]

The manufacture of the Hotchkiss hammer was eventually moved to Brooklyn, New York, where two of Charles Merrill's sons established their own firm in 1866. That company came to

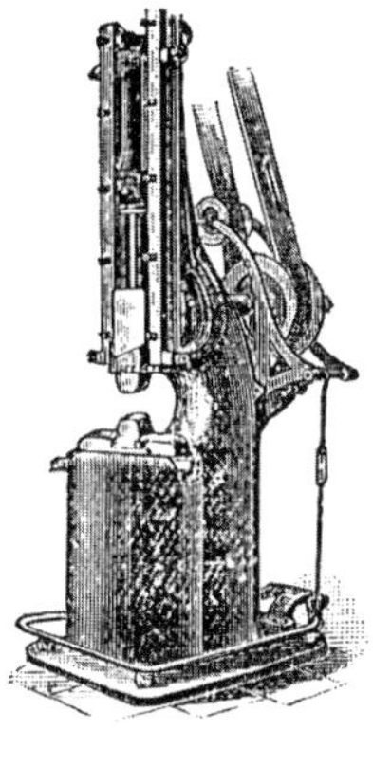

The Hotchkiss Hammer became the Merrill Upright Power Trip Hammer. This cut is from 1892. (*The Journal of Railway Appliances and Railway Price Current*)

specialize in the building of board drop hammers and the production of drop forgings. In 1905, the company moved to Maspeth, New York; and a listing in the 1909 *Thomas' Register of American Manufacturers* still mentioned an "air-cushioned trip" hammer in its line.[31] Since Merrill Brothers built drop hammers for many years thereafter, it is difficult to determine precisely when the Hotchkiss hammer was finally discontinued.

The Hotchkiss hammer was favored by some of America's top machine shops during the nineteenth century. For example, Pratt and Whitney of Hartford, Connecticut, employed a 60-pound Hotchkiss Atmospheric Hammer alongside a 700-pound Marchand and Morgan steam hammer, three drop hammers, and an adjustable-stroke trip hammer (of their own manufacture).[32] Also, an engraving of the forge shop of the Singer Sewing Machine Company, in Elizabethport, New Jersey (ca. 1880), provides a rare glimpse of a Hotchkiss hammer in use.[33]

The unmistakable form of a Hotchkiss Atmospheric Hammer at work in the Singer Sewing Machine Company drop-forging plant, in Elizabethport, New Jersey, ca. 1880. (*Genius Rewarded; or the Story of the Sewing Machine*)

* * *

The story of crank-actuated open-die power forging hammers thus began in the United States during the 1850's. American inventors devised ingenious alternatives to the expensive and slow direct-acting steam hammer. The so-called "atmospheric" hammer of Bernard Hughes was capable of rapid blows and provided subtle control to the blacksmith. Furthermore, it ran off of a common lineshaft, and was affordable. Hughes' hammer was trip-actuated, but it was not long before Milo Peck applied his experience with crank-actuated drops to propose a crank-actuated atmospheric hammer. Peck's invention was quickly followed by that of Bennet Hotchkiss, whose simplified atmospheric hammer was subsequently manufactured for over half a century. Hotchkiss was the first to use a spring (air-cylinder) linkage in between a crank and a hammer ram. Metal springs were adapted in lieu of the air cylinder immediately after the Civil War. With this subtle improvement, the widespread use and development of crank-actuated power forging hammers burgeoned. However, the sequence of this important chapter in American mechanical invention was soon forgotten. In his detailed report for the Tenth Census on *Machine-tools and Wood-working Machinery* (1885), F. R. Hutton introduced Chenot's (self-contained) Atmospheric Hammer by remarking that it was "a very natural step to replace the elasticity of a steel spring by the elasticity of a cushion of air and apply it to produce dead-blows with a crank hammer."[34]

In the chapters which follow, the descendants of these American atmospheric hammers are organized according to geometry (helve or linear guided-ram) and by the location of the actuating crank (vertically or non-vertically-configured). As explained, these divisions are somewhat arbitrary, and by no means do they separate all hammer-building companies cleanly. Just as Milo Peck noted that his atmospheric hammer could have its crank either above or behind the ram, some hammers were, in fact, built in both styles. A number of companies built hammers in two of these categories, and one of them, the C. C. Bradley Company, built machines in all three. Crank-actuated helves will be considered first, since the venerable helve hammer pre-dates those machines using a linear motion in guided-rams. Hammers with a vertically-configured guided-ram mechanism will follow, after which will come those machines with a non-vertical configuration.

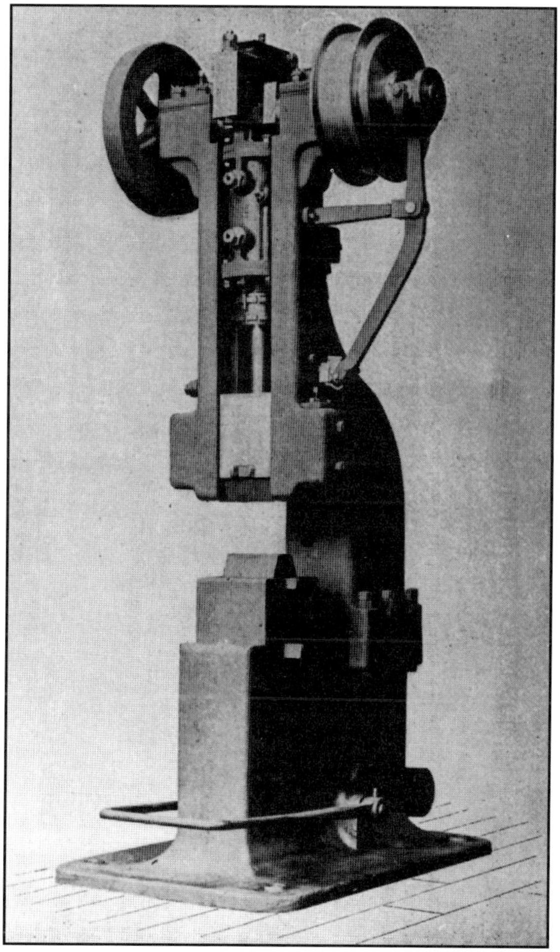

Three decades after the first appearance of crank-actuated atmospheric hammers in the United States, the air-cylinder spring linkage was re-introduced from England. The vertically-configured Hackney Pneumatic Power Hammer will be discussed in Chapter 4.

Paye's Hammer, the first crank-actuated helve, was a sophisticated machine. The stroke could be readily adjusted by means of a hand-wheel, even while the hammer was in operation. (*The Scientific American*)

CHAPTER 3

"The Favorite of Every Hammerman"

The first crank-actuated helve hammer was designed by Edward Paye, of New York City. Paye was granted patent number 30,343 on October 9, 1860; and the rights were assigned to Cornelius Delemater. The hammer was soon placed into production at the Delemater Iron Works. At the time of its description in *The Scientific American,* in July of 1861, the hammer was being used in two other important engineering establishments in New York City: the Allaire Works as well as Fletcher and Harrisons. It was claimed that this hammer was "adapted to all work, great and small, of a steam engine manufactory or other iron works"[1]

Paye's hammer featured a clever means of enhancing and varying the stroke. Its belt-driven crank oscillated a lever which was connected to the helve by a yielding connecting rod. This connecting rod, or pitman, was composed of a rod and cylinder, which contained a coil spring. The spring was compressed on the up-stroke, and the energy thus stored increased the force of the blow. The stroke of the hammer could be adjusted by moving the connecting rod along the curved arm of the oscillating lever. This was easily accomplished by turning a handwheel, with the hammer

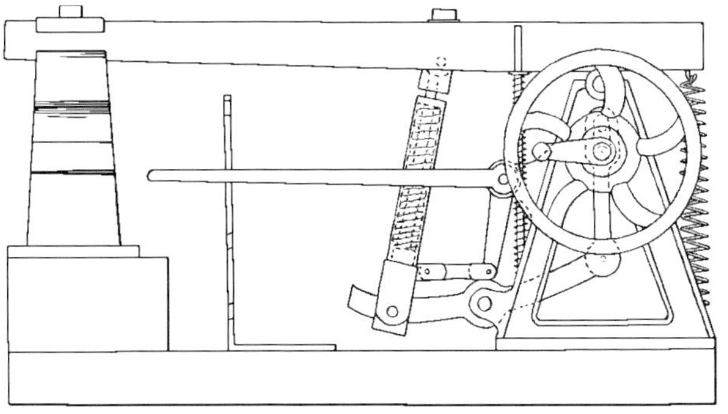

In Edward Paye's patent of 1860, the variable stroke was adjusted by means of a lever.

either idle or in full operation. Springs placed vertically on either side of the fulcrum helped to balance the helve, and to hold the hammer open between operations. An idler pulley and friction brake were used to start and stop the hammer. Paye claimed that his machine effectively took the place of three ordinary hammers, and that "the design would do a given quantity of work with one fourth of the power required for the common tilt."[2]

The Delemater Iron Works, where Paye's hammer was built, was one of the most important American shops of its day. Cornelius Delemater had become a clerk in the Phoenix Iron Works at the age of twenty one. When the firm dissolved four years later, Delemater and a partner continued the business. In 1850, the operation was moved uptown, to the foot of 13th Street; and, with the retirement of his partner, the company became the Delemater Iron Works. The enterprise grew steadily, encompassing the largest branches of mechanical construction. In the Fall of 1861, soon after the appearances of Paye's hammer, Cornelius Delemater executed a historic commission: the engines for the U.S.S. Monitor. These twin "vibrating lever" engines were designed by the Monitor's eccentric inventor, John Ericsson. When the Monitor steamed to Hampton Roads, and its famous rendezvous with the Confederate Ironclad Merrimac, her boiler and engines were manned by Delemater workmen.[3] It is quite possible that forge hammers of Paye's advanced design played a role in the execution of this vital wartime commission. Over the years, the Delemater Iron Works worked closely with John Ericsson, constructing his gunboat, Destroyer, as well as many of the earliest screw propellers. In 1870, the Delemater Iron Works produced a fleet of thirty gunboats for the Spanish government. Each of these vessels was 110 feet long, capable of ten knots, and was armed with a 100-pound Parrott rifle. The entire commission was completed in four months! As a further indication of the scale of work at this one establishment, by the time of Cornelius Delemater's death in 1889, the firm had produced 251 marine engines.[4] The extent and duration of the manufacture of Paye's hammer amidst this activity is unknown, but the demand for forgings in this one shop alone must have been considerable.

*   *   *

The development of the crank-actuated helve hammer proceeded slowly in the 1860's. Hotchkiss' Atmospheric Hammer had gained a substantial following in the North during the Civil War. Immediately after the war an even more successful variation on the theme appeared in the form of the Shaw and Justice Dead Stroke Power Hammer, with its semi-elliptical metallic spring linkage. Machines of Shaw's design were built with the crank positioned both directly above the ram (vertically-configured) and with the crank behind the ram, working a pivoting beam (non-vertically-configured). Each of these configurations was extensively developed in the decades which followed. At the same time, the ancient trip-actuated helve remained in use at many forges in America. Trip-actuated helve hammers persisted as an economical alternative to the new state-of-the-art, and they continued to be manufactured by such important tool builders as Pratt and Whitney of Hartford, Connecticut.

Despite this competition, several American inventors followed Paye's lead in pursuing a crank-actuated helve. Alfred J. Grainger, of Wilmington, Illinois, was granted patent number 62,407 on February 26, 1867, for his design. Grainger's hammer used a helve which was made up of two in-line sections. These sections were held together at a pivoting joint by flat steel springs. The crank drive mechanism was thus isolated from the shock of the hammer's impact; and, like Paye's spring-cylinder connecting rod, stored energy for an enhanced blow. The crank included a counter-weight to maintain smooth running of the machine, and to hold the hammer open between operations. It is not known if this hammer was ever manufactured; however, the design proved influential. Within several years, in nearby Chicago, the groundwork was laid for another crank-actuated helve design, one which was to dominate an important share of the power forging hammer market for more than seventy years.

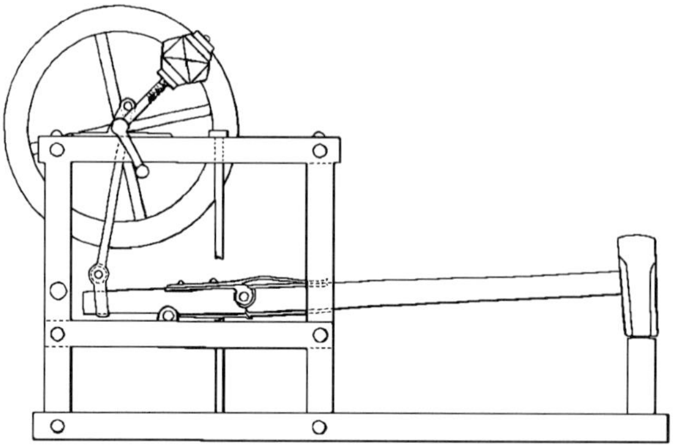

In Alfred Grainger's patent of 1867, forward and rear sections of the helve were connected flexibly by flat steel springs.

Treat T. Prosser, of Chicago, Illinois, was granted patent number 88,980 on April 13, 1869, for a clever arrangement that also used a two-part helve. However, in Prosser's design, the portion of the helve which was oscillated by the crank pivoted separately below the helve, connected to it by coil springs. Once again, these springs provided isolation from shock and stored energy to yield a blow far stronger than would have been yielded by gravity alone. The stroke was varied by increasing or decreasing the speed of the crank through a foot-controlled friction drive. Despite the probable limitations of this last feature, the application of an independently-oscillating beam to work the helve was inspired.

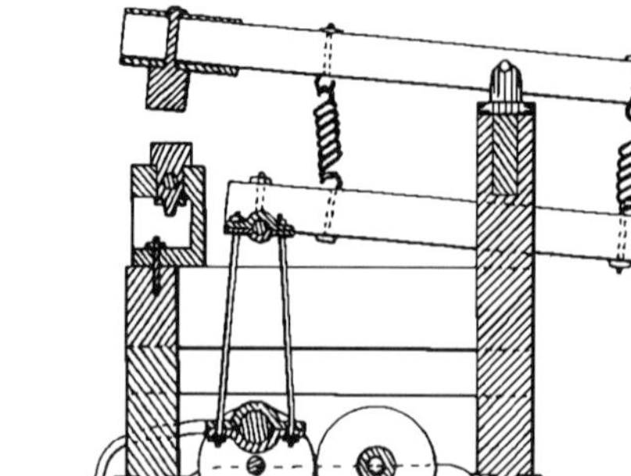

In Treat Prosser's patent of 1869, the crank oscillated a pivoting beam which was connected to the helve by coil springs.

John C. Butterfield and James Hay, both of Chicago, soon patented a hammer design which included a spring-cushioned oscillator much like that of Treat Prosser. The Butterfield and Hay design, which was granted patent number 108,326 on October 18, 1870, differed in that rubber cushions were used in lieu of coil springs, and a more sophisticated means of adjusting the stroke was provided. The length of the rod which connected the crank to the oscillator could be changed to accommodate work of varying thicknesses, and an adjustable eccentric crank varied the intensity of the stroke. The development of this hammer was likely affected by the

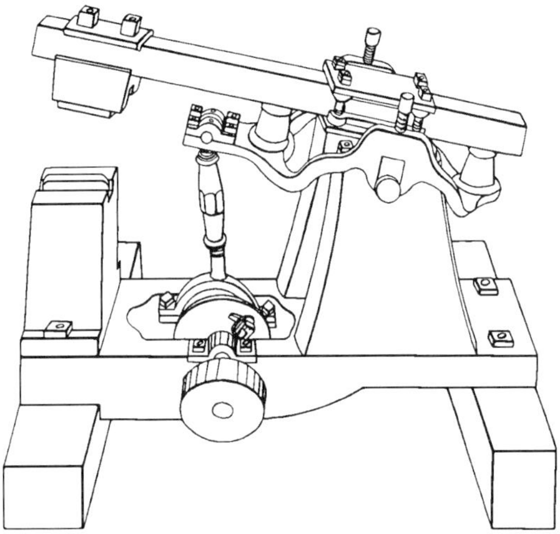

John Butterfield and James Hay's patent of 1870 was similar to Prosser's, but used rubber cushions for the connection between the helve and the rocking carriage.

disastrous fire that consumed Chicago in 1871. In any event, the rights to the design were soon acquired by a company in Syracuse, New York, where production of the hammer began in earnest.

That machine, the Bradley Rubber Cushioned Helve Hammer, occupies a remarkable place in the history of forging machinery. As a helve hammer its lineage can be traced back to the Middle Ages in Europe, and to the mid-seventeenth century in America. But the Bradley helve also bridged the worlds of the nineteenth-century blacksmith-mechanic and the twentieth-century precision machinist. As industry moved inexorably away from skill-intensive (but less precise) "smith" forging methods, Bradley promoted the versatility of its hammers and their capability of holding tolerances as fine as one thousandth of an inch for certain forgings. Introduced in 1872, the Bradley Rubber Cushioned Helve Hammer was an immediate success. It won recognition at the industrial fairs of its day, stimulated earnest competition, and brought continuing prosperity to a company that was already four decades old.

* * *

Christopher Columbus Bradley arrived in Cossit's Corners, New York, in 1822. At the time, there was little to commend the site beyond its location on the soon-to-be-completed Erie Canal. But when the Canal opened in 1825, everything changed. Travel time from New York to Buffalo was cut from three weeks to six days, and freight costs dropped ninety percent. The sleepy hamlet grew to become the city of Syracuse. C.C. Bradley and his descendants played an important role in that growth. In 1832, C.C. Bradley and a partner opened the first foundry in Syracuse. Alexander, Bradley & Company was a general foundry, with a lucrative specialty in the casting and repair of salt kettles.[5] Two years after the foundry opened, the second Christopher Columbus Bradley was born. He grew up amidst the sand and smoke of the foundry, and joined the business in his mid-teens. In 1855, Bradley (now a sole proprietorship) moved from the "Old City Foundry" to new quarters at the corner of Wyoming and Marcellus Streets. Bradley, Sr. took his sons, C. C. Jr. and Waterman Chapman Bradley into the partnership, and the company became C. C. Bradley & Sons. With the subsequent withdrawal of W. C. Bradley, the firm became C. C. Bradley & Son. The elder C. C. Bradley died in 1872; and, at that point, the firm became simply: Bradley & Company.

By that time, Bradley had successfully expanded its foundry business, and was prominent in the manufacture of carriages and agricultural implements. Its shops now included facilities for the

The popular Bradley Handy Wagon cost about $70 in the late 1880's. (*Bradley Company trade literature*, from the collections of the Onondaga Historical Association)

woodwork, machine work and forging necessary for these specialties. Taking full advantage of its excellent location, with sales representatives in New York City and Boston, Bradley produced a full line of vehicles including buckboards, buggies, phaetons, road carts, and surreys. Its output in agricultural implements was no less impressive, offering assorted harvesters, mowers, reapers and rakes. To remain at the forefront of a rapidly changing technology, Bradley licensed and manufactured promising designs developed outside the company. M.G. Hubbard's Patent Harvester, for example, was produced at the time of the Civil War. Despite the eventual popularity of its hammers, and the enormous volume of work which their manufacture entailed, Bradley remained prominent in the transportation and agricultural fields well into the twentieth century.

As manufacturers of the diverse iron hardware required for carriages, the forge shop at Bradley would have been acutely aware of the shortcomings of the power hammers of its day. Bradley management would also have been well-qualified to assess the need for a versatile, high quality hammer and its potential market. The results upon its introduction of the Butterfield and Hay design were surely gratifying. *The Scientific American* appreciated the merits of the rubber-cushioned suspension:

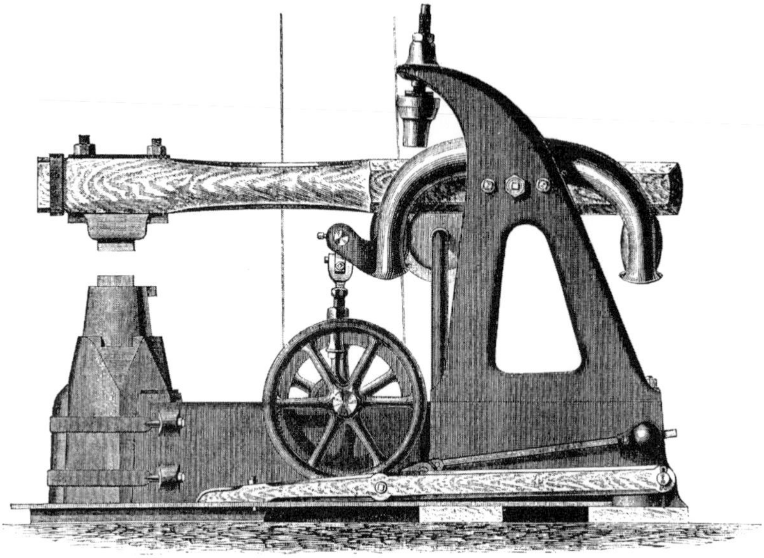

An early illustration of the Bradley Rubber Cushioned Hammer from 1873. (*The Scientific American*)

"The force and power of the blow is greatly influenced by the reactive and united action of the cushions. So harmonious is this combined action upon the motion of the helve, that it is stated that an observer holding his hand upon the working parts when under the most rapid and violent movement can hardly identify the strokes of the hammer."[6]

An English engineer touring America viewed one of the early Bradley hammers and reported that it ran with a "stillness and charm" possessed by neither the Shaw & Justice hammer nor the competing Hotchkiss Atmospheric Hammer.[7] After little more than a year on the market, Bradley boasted, "It is the favorite of every hammerman; it promptly obeys his every touch, and he soon regards it as a thing of life."[8]

John C. Butterfield remained involved in the refinement of the Bradley helve. He was granted two consecutive patents on October 28, 1873, which further reveal the developmental process. The first, patent number 144,058, dealt with a problem that occurred in the larger sizes of the hammer. It was found practically impossible to keep the hammer from rebounding and striking a

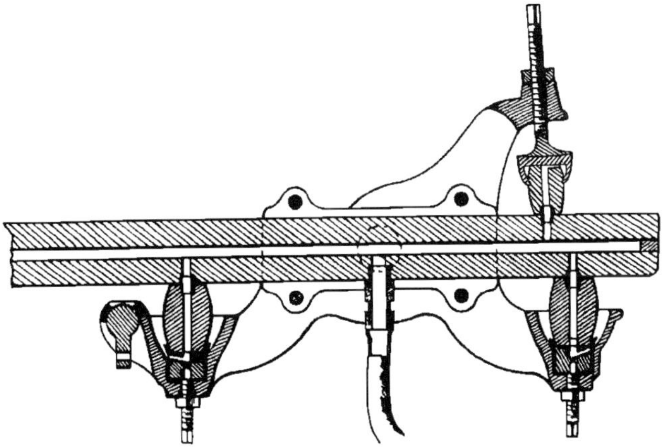

John Butterfield's first patent from 1873 included the addition of a rubber cushion above the helve, and an air-blast (through a hole bored in the helve) to keep the cushions from over-heating.

second blow on every down stroke. Since there were only two rubber cushions (both below the helve), the rebound occurred before the forward cushion could begin to lift the helve. To stabilize the helve between blows, Butterfield proposed placing additional cushioning above the helve. He also addressed concerns for the independent adjustment of the rubber cushions, and outlined a fascinating solution to the problem of overheating encountered when the rubber cushions were run at full speed. A blast of cool air could be channeled to each rubber cushion through a hole bored the length of the helve. Butterfield thought that such a hole might also enhance the stability of the wood as it cured. It is not known if this last improvement was ever incorporated into actual hammers.

The second Butterfield patent of 1873, number 144,059, encompassed many of the salient design details of the Bradley hammer as it first appeared in industrial trade journals. To contend with the problem of rebound, the design now included a rubber cushion affixed to the hammer frame forward of the pivot. The patent also detailed the adjustable cross-head (at the point where the helve pivoted), the eccentric by which the throw was adjusted, and a universal joint between the eccentric rod and the carriage. It is not known if John C. Butterfield made any further contributions

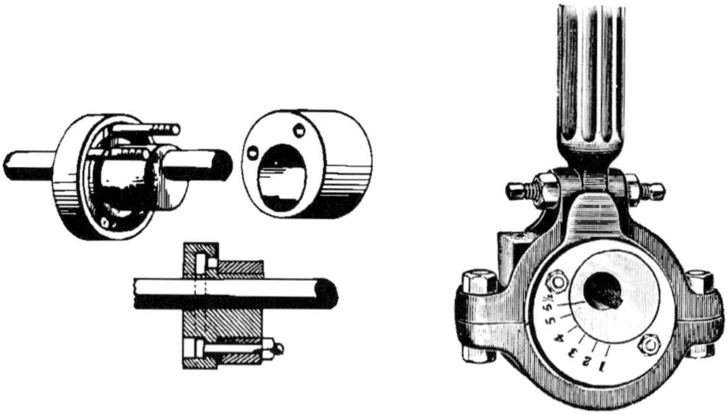

John Butterfield's second patent of 1873 included a description of the adjustable eccentric. The illustration of this feature at the right comes from a Bradley trade catalogue from half a century later.

to the Bradley hammer line, but his professional interest in power forging hammers continued. A decade of subsequent patents reveal that Butterfield became interested in hammers cushioned by air cylinders. Although the actual manufacture of Butterfield's atmospheric hammers has not been confirmed by the author, it seems likely. Since these patented designs were for guided-ram hammers of non-vertical configuration, they will be described in Chapter 5.

When the Bradley Rubber Cushioned Helve Hammer appeared on the market, vulcanized rubber was still an exotic engineering material. In 1844, Charles Goodyear patented the process whereby raw rubber was heated with sulphur and made stable. Initial applications included footwear, waterproofing of garments, fire hose, rubber bands, door springs and inflatable beds. But it was not until the bicycle craze swept America in the late 1880's and 1890's that rubber came into its own as the premier material for tires. Despite this, a dependable supply of natural rubber from plantations on the Malay Peninsula did not begin until 1907. Up to that time, rubber was harvested from wild rubber trees, and the supply was less than certain.[9] The use of rubber cushions in a heavy, industrial machine was greeted with some skepticism. A

British observer at the Centennial Exhibition in Philadelphia, described with skepticism the india-rubber cushions of the Bradley hammer for the readers of *Engineering* magazine. "Not having any fancy for this material as one for machine parts," he illustrated the principle of the Bradley hammer with steel leaf springs![10]

The Bradley Rubber Cushioned Helve Hammer at the Centennial Exhibition. (*Masterpieces of the Centennial Exhibition*)

By the time of the 1876 Centennial Exhibition, the Bradley Rubber Cushioned Helve Hammer had incorporated Butterfield's improvements and assumed its mature form. The oscillating carriage included an upper cushion behind the pivot to better control the upward movement of the helve. Another cushion was positioned on the upper hammer frame, forward of the pivot, to arrest the helve at the top of its stroke. Though several variations were eventually added to the Bradley line, the rubber cushioned helve remained in production, relatively unchanged, for more than three quarters of a century. In order to accommodate the diverse manufacturing niches that might require its hammer, Bradley built the Rubber Cushioned Helve in a variety of sizes. Its smallest version had a ram weighing a mere fifteen pounds. Running at 435 blows per minute, it weighed 1,350 pounds overall, and was suited to work averaging three quarters of an inch in diameter. The smaller hammers were ideally suited to the demanding work of cutlery manufacture; and in 1881, Bradley promoted its line by sending out pocket knife blades forged under a 40-pound helve.[11] The largest model boasted a ram weighing 200 pounds. Running at about 230 blows per minute, it weighed 9,500 pounds overall and could handle work three and a half to four inches in diameter. Between these extremes were more than half a dozen sizes with special configurations available for special tasks. For example, in the mid-1880s, special 40-, 60- and 70-pound hammers were available for forging hay forks, and special short-bed 70- and 100-pound hammers were offered for forging plate.[12]

Bradley hammers were not cheap. A manufacturer's tool par excellence, they were never really intended for the village blacksmith. A 100-pound Bradley hammer purchased through Manning, Maxwell & Moore of New York, in 1884, cost a substantial $850. A comparable cushioned helve hammer built by one of Bradley's primary competitors, the Long & Allstatter Company, of Hamilton, Ohio, was a bit more at $875;[13] but, there were many less expensive helve hammers on the market. A 100-pound guided-ram "helve" from Jenkins & Lingle, of Bellefonte, PA, could be purchased from the manufacturer for just $450 (1885). The popular Dead Stroke Power Hammer built by Shaw and Justice fell between these figures, with a 100-pound hammer costing $680 (1884).[14] For forgings at the larger end of the Bradley's capacity there was even competition from the smaller steam hammers of the day. A 6" steam hammer built by the firm of Richard Dudgeon, then located in New York City, cost $1,050 (1881), about the same price as the 200-pound Bradley helve.[15] During the first decades of the twentieth century, the competition for the general blacksmith's market in power hammers became particularly fierce. Economically-priced hammers such as the 30-pound Kerrihard, from Red Oak, Iowa, could be purchased for as little as $60 (1909).[16] A comparable Bradley cost five times that amount.[17]

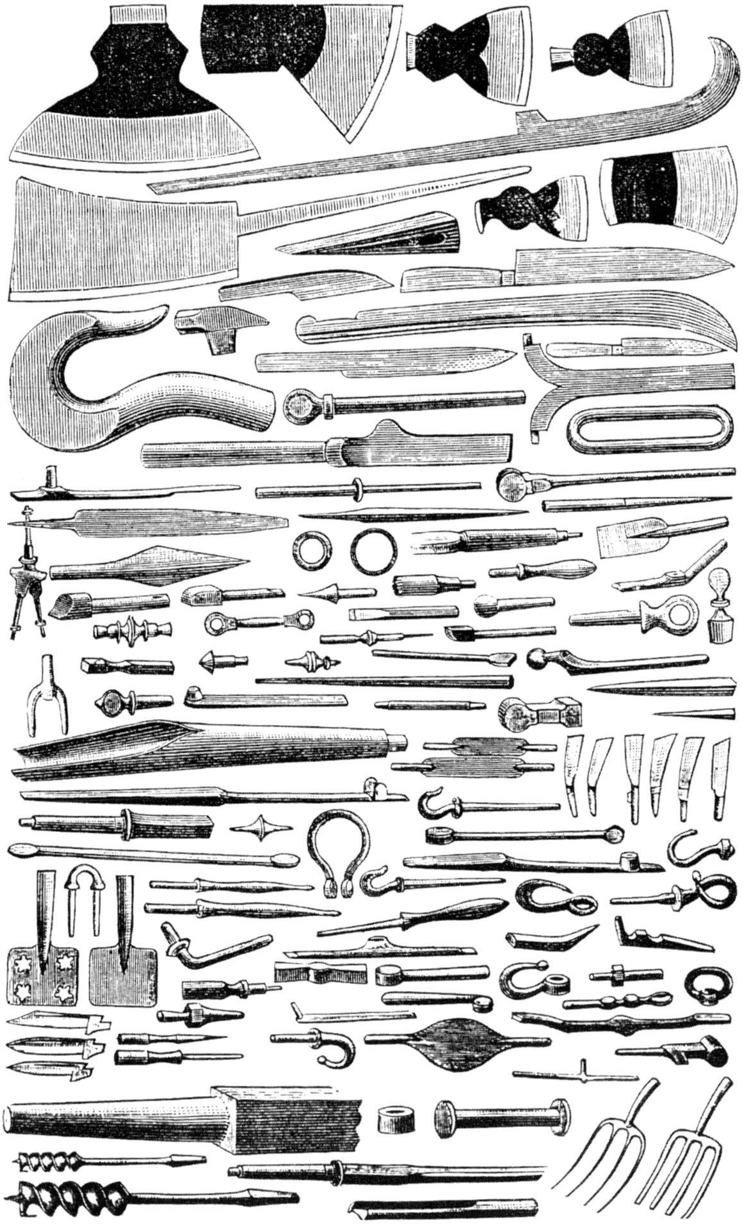

A selection of forgings done with the Bradley Rubber Cushioned Helve Hammer ca. 1881. It was observed that, "the use of hammers of this class has had the effect to cheapen enormously the production of these and similar classes of forgings." (*American Machinist*)

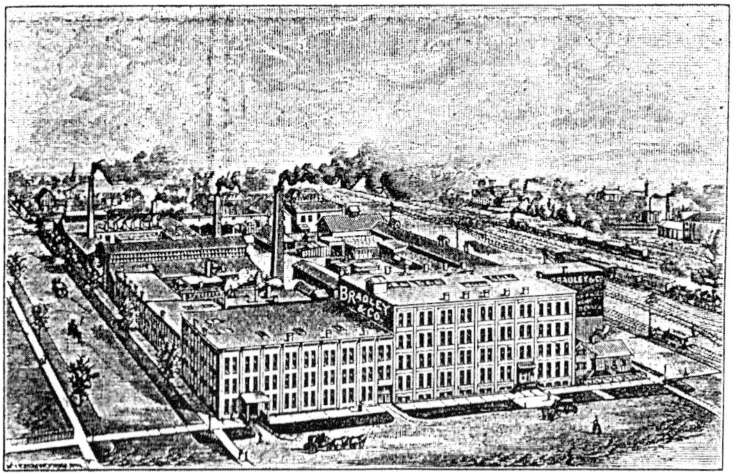

A view of the Bradley factory ca. 1888. (*Syracuse, N.Y. Illustrated,* from the collections of the Onondaga Historical Association, Syracuse)

Having received the Grand Medal of Merit for their hammer at the Centennial Exhibition, orders multiplied rapidly. This success was all the more impressive given the chilling effects of a six-year depression which began in 1873. After nearly half a century (1878), the company finally incorporated in the state of New York. In 1879, Bradley reported to the *American Machinist* that its workforce of 150 was extremely busy, and had just shipped a hammer to Manchester, England.[18] Bradley was an important factor in the economy of Syracuse, with its machinists bringing home as much as $2.50 a day, blacksmiths and woodworkers $2.25, and helpers $1.00.[19] Sales doubled in 1880, and with an expanded workforce of 170, Bradley enlarged its plant.[20] At times, the factory produced an average of one hammer a day, and often ran on overtime. By mid-1882, the workforce numbered more than 200, and Bradley proudly announced that there were more than 700 Bradley hammers in use.[21] By 1887, Bradley employed 260 men.[22] In 1888, Bradley had sales of about $400,000 and the factory, with its three 150-horsepower engines, was a Syracuse landmark.[23] By the end of the decade, there were over 1,500 Bradley hammers in use.[24]

With success in the power hammer market, and the continuing prosperity of the vehicle and agricultural implement divisions, the Bradley company was able to pursue an aggressive program of research and development. In the early 1880s, the company introduced the Bradley Hard Coal Heating Forge. It was advertised as the perfect solution to the bottleneck created by the

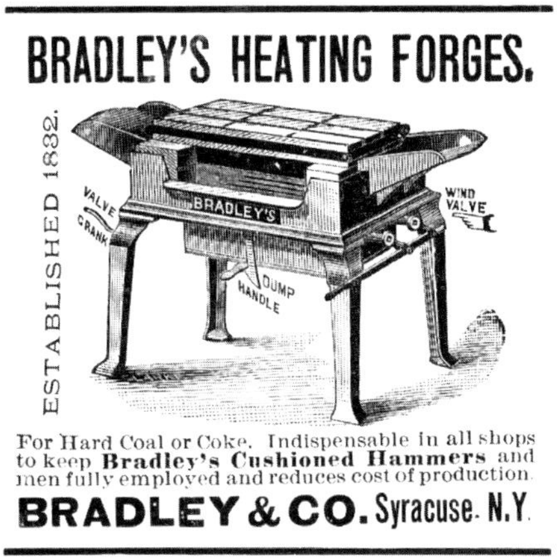

The Bradley Hard Coal Heating Forge was introduced to complement the Bradley hammer line, reducing the bottleneck created by such efficient forging machinery. (*American Machinist*)

accelerated volume of work that could be produced with Bradley hammer.[25] Facing competition from the Beaudry Company of Boston, Massachusetts, Bradley developed a line of guided-ram "helves," in the mid-1880's. These are discussed in the section on non-vertically configured, guided-ram hammers. Just before the turn of the century, Bradley also entered the realm of vertically-configured, guided-ram hammers with the introduction of its so-called Compact Hammer. In addition to the numerous patents pertaining to these non-helve Bradley hammers, there were several later patents concerned with the Rubber Cushioned Helve Hammer. Herman J. Bierhart, of Syracuse, was granted patent number 385,540 on July 3, 1888. It described an adjustable anvil block for maintaining the exact alignment of dies. Christopher Columbus Bradley himself was granted patent number 497,012 on May 9, 1893. This patent protected a ball-joint intended to eliminate the harmful effects arising from twisting or misalignment between the eccentric drive and the oscillating carriage. This sort of engineering detail was of particular interest to C. C. Bradley, Jr., who also developed the highly successful Bradley Ball Bearing Shaft Coupler for the carriage trade. So popular was this device that when a pundit claimed "the Bradley couplers are known from pole

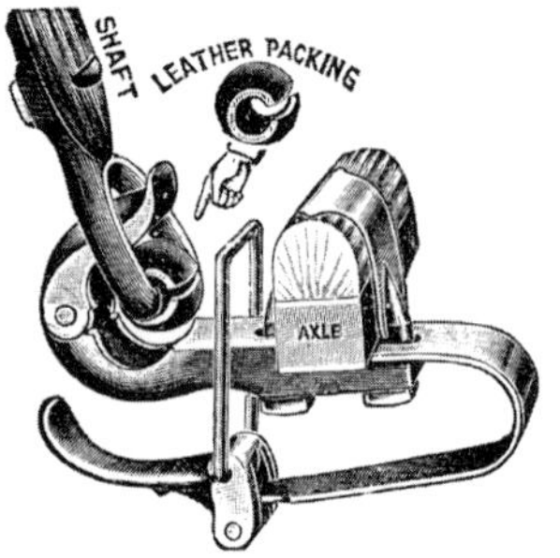

Christopher Columbus Bradley (1834-1916) was President of the company when the first Rubber Cushioned Helve was built. He also contributed the design of a popular coupler for carriage shafts. (*The Blacksmith & Wheelwright*)

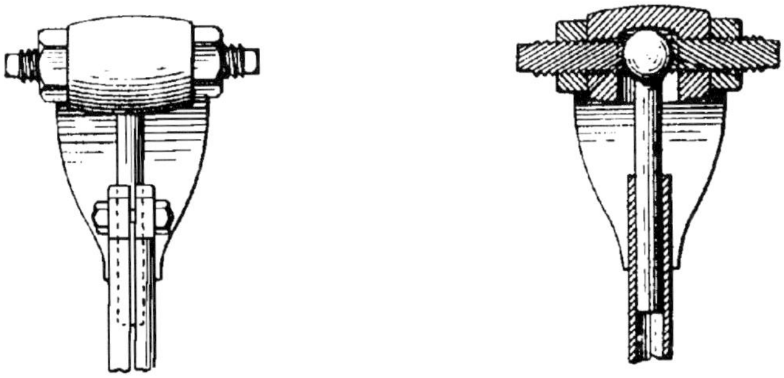

Akin to C.C. Bradley's shaft coupler was his 1893 patent for a ball-joint connection between a power hammer's pitman and the helve.

to pole," it was close to the truth![26] Also included in Bradley's 1893 patent was a novel tensioning mechanism for the leather strap in the Bradley Upright Strap Hammer (described at length in the chapter on non-vertical guided-ram hammers). While this later improvement was incorporated into actual hammers (briefly), use of the ball-joint is not known.

In the mid-1890s the Bradley enterprise was restructured. A separate Bradley Company was incorporated in 1894, specifically to attend to the manufacture of the hammers and forges. Another company, C.C. Bradley & Son, Inc. was formed in 1896, for the manufacture and sales of the carriage shaft couplers. Despite the company name, the vice president of this branch was Cora M. Bradley, a daughter. In 1902 (upon the death of Waterman Chapman Bradley, who had served as secretary and treasurer of the Bradley Company), the firm was reunited. The Bradley Company was absorbed by C.C. Bradley & Son, Inc. The second Christopher Columbus Bradley retired from an active role in the company in about 1911, and died in 1916.[27] The reins were taken-up smoothly by his son, know as C.C. Bradley, Jr., who had joined the company more than twenty-five years earlier. Born in the year following the introduction of the Rubber Cushioned Helve Hammer, he studied machine-shop work for two years at the Chicago Manual Training School. His plans to attend Cornell University were thwarted by illness, and he joined the family business. He became vice president of the Bradley Company in 1894.

The first decade of the twentieth century brought intense competition for the Bradley hammers. New materials and processes challenged the place of open-die, ferrous forgings in manufacturing. Also, the sudden expansion of the power hammer market into the realm of the small-scale, repair-oriented blacksmith gave rise to a new generation of helve hammers. Occasionally, as in the case of the Rochester Helve Hammer (built by the West Tire Setter Company, of Rochester, New York), these hammers evolved to provide competition to Bradley in the manufacturing realm as well. The helve hammer proved to be a surprisingly durable mechanical design. Even in the mid-1920's, the Beaudry Company, of Boston, Massachusetts, attempted to introduce a new air-cushioned helve hammer. The chief competition, though, came from guided-ram hammers and self-contained, pneumatic machines. The Bradley company responded with an advertising campaign that stressed the longevity and dependability of its hammers, and offered highly detailed examples of their successes. One advertisement, for example, described the production of line bar forgings (straight pry bars) at Hubbard and Company of Pittsburgh: using an eight-year-old, 300-pound Bradley Upright Hammer, 1.25 inch square stock (of seventy five point carbon steel) was drawn-out in one heat. The operation consisted of forging 35 inches to a circular cross-section (slightly tapered), 15 inches to hexagonal, leaving the 14 inch balance as is. One thousand of these bars were produced each day![28] Another advertisement described the use of Bradley hammers at the American Fork and Hoe Company. Having purchased the very hammer displayed by Bradley at the Centennial Exhibition in 1876, the company continued to employ Bradley helve hammers after more than forty years. Much of the work performed on its machines was the drawing-out of stock, as in forging the tines for hay forks. For one agricultural tool, a piece of steel three inches long was drawn out to 24 inches.[29]

In a series of "New Bradley Developments," the company distributed details of further capabilities and accomplishments of its hammers. Applications involving special dies were especially featured. In one instance, a Bradley hammer was used for the cold restriking of a drop forging. In this case, the drop forging process alone could not hold the necessary tolerances on a spherical section, but the restriking process, between fairly simple dies, brought the part to within the required two thousandths of a inch and eliminated the need for costly machining.[30] Another unusual cold operation performed on Bradley hammers was the forming of special, stainless steel rods used in the door locking assembly of streamline trains. The operation was performed cold to preserve the properties of the steel, and the dies cost only $45 to $100 per pair.[31] Yet another example included the tapering of automobile torque tubes. Despite the development of specialized forging technology, such as rotary swaging, a Bradley hammer, fitted with relatively inexpensive dies remained competitive.

C.C. Bradley & Son, Inc. weathered the Great Depression, but financial worries were the likely cause of tragedy for the firm. On March 9, 1932, in the punch press department of his factory, Christopher Columbus Bradley, Jr. shot himself. Several weeks before, an equity receivership had been granted in Federal Court.[32] The action was instituted against Bradley by the Sawhill Manufacturing Company, of Pittsburgh, a creditor claiming $5,000. Bradley was succeeded by his sons: Charles Goodwin Bradley and yet another Christopher Columbus Bradley. The years following World War II were rough for American machinery builders; and in 1949, C.C. Bradley and Son, Inc. merged with the more prosperous Edlund Machinery Co., of Cortland, New York. As a division of the new Bradley-Edlund Corporation, the hammer-building operation was moved to Cortland in 1952.

"Bradley Hammers"

The Edlund Machinery Co. also had its beginnings in Syracuse. John A. Edlund, a Swedish immigrant, established a general machinery-building firm with a partner in 1912. The company moved to Cortland in 1914; and in 1918, it began to concentrate on drilling machines. Its multi-spindle drills became a standard among manufacturers in the automotive and aviation fields. These machines were of immense importance in the war effort; and by 1941, Edlund employed 180 workers building 30 types of drill.[33] In the decade following World War II, the company consolidated to less than half that number of workers, but remained at the forefront of its machinery-building niche. During the 1950's, control of the company, and the Bradley hammers, was acquired by Precision Castings Company of Fayetteville and Syracuse. In turn, Precision Castings Company became a division of Harsco Corporation, of Harrisburg, Pennsylvania, an international manufacturing organization with nearly twenty plants in the United States and many overseas. While hammers continued to be built and serviced, the Bradley company name passed out of use on July 29, 1955, with the formal dissolution of the Bradley-Edlund Corporation. Harsco's Edlund Machinery Co. Division remained at the cutting-edge of the drilling machine field, and eventually it attracted another suitor.

In 1960, Edlund unveiled its new, $35,000 numerically-controlled drilling, tapping and boring machine. This marvel boasted a General Electric electronic control system programmed by paper-tape.[34] Edlund's state-of-the-art approach to drilling machines and its developments in automatic tool-changers attracted the attention of the Monarch Machine Tool Company, of Sidney, Ohio. Monarch was one of the foremost lathe builders in the United States, and it bought the Edlund Machinery Company in 1963. Along with the engineering and manufacturing assets of Edlund, Monarch also acquired the dwindling Bradley hammer line. Monarch offered service and support for Bradley hammers into the early 1970's, at which time the existing patterns and drawings were acquired by the Cortland Machine and Tool Company, Inc. This firm had previously subcontracted to Edlund and Monarch, making parts for the hammers. The Cortland Machine and Tool Company, Inc., still possesses hammer drawings dating back to the early 1900's, as well as some patterns. Despite a steady decline in the call for support of the Bradley hammer line, the company continues to offer some services.[35]

* * *

The Long and Allstatter Company of Hamilton, Ohio, was one of the Bradley Company's earliest competitors in the field of crank-actuated helve hammers. Like the Bradley hammer, the Long and Allstatter Adjustable Cushioned Helve Hammer used rubber cushions to isolate the helve from the drive mechanism. However, the Long and Allstatter machine was designed to be readily adjustable for a wide range of work thicknesses. This was made possible by placing the helve and the pivoting yoke (which supported the helve between rubber cushions), as well as the crank and drive shaft, onto a frame which could be adjusted vertically. The frame was suspended from threaded rods on either side of the helve, and an arrangement of bevel gears allowed the frame to be

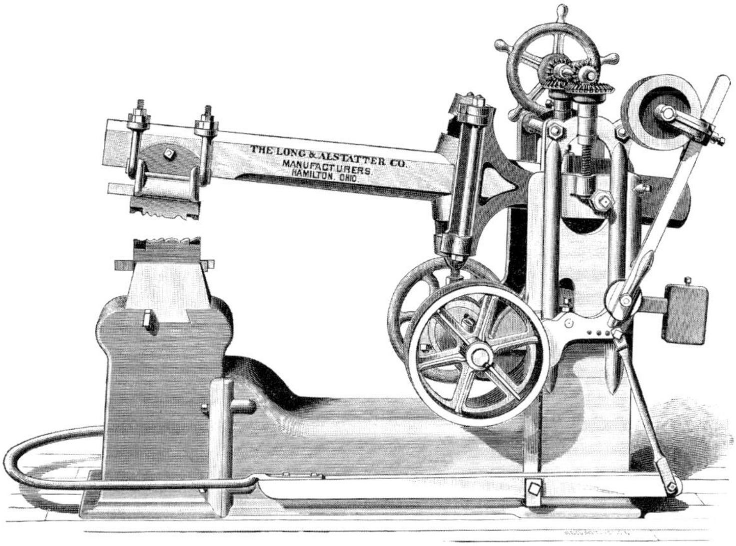

The Long and Allstatter helve soon provided competition for the Bradley hammer. (*Long & Allstatter Company catalogue*)

smoothly raised or lowered by turning a single handwheel. This facilitated working with various thicknesses of stock, but it also allowed the hammerman to readily interchange and use dies of different thicknesses without requiring painstaking shimming (a great convenience). Additionally, the actuating crank of the Long and Allstatter Hammer included an adjustable cam, which provided for variable blows. For these innovations, John M. Long was granted patent number 136,253 on February 25, 1873.

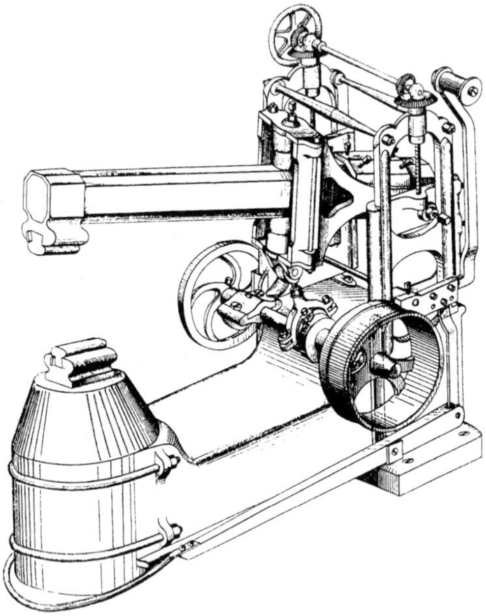

John M. Long's patent of 1873 described the hand-wheel and bevel-gear arrangement used to raise and lower the helve.

The Long and Allstatter Company was probably best known for its power punches and shears. The enterprise began in 1856, with the formation of Long, Black and Allstatter. John M. Long, Peter Black and Robert Allstatter prospered in their business, which combined general smithing with some machine work. In 1867 the company embarked upon the manufacture of reapers, and grew rapidly. Upon the retirement of Peter Black, in 1871, the firm became the Long and Allstatter Company. Its helve hammer was patented at a time of particularly rapid growth, and in that year the company also moved into larger quarters. While the company rose to great prominence in the field of punches and shears, its agricultural implement department continued to develop and manufacture cultivators, disc harrows, as well as moldboard and disc plows, well into the twentieth century.[36] As hammer builders, the Long and Allstatter Company was conservative. The Adjustable Cushioned Helve Hammer was manufactured for more than sixty years, with only one substantial revision. This occurred at about the time of the First World War. Also, at that point, the hammer was patriotically re-named the American Rubber Cushioned Helve Hammer, and the plant became known as the American Punch and Shear Works.

Initially referred to as the Improved Adjustable-Helve, Elastic Stroke Power-Hammer, the earliest machines were built in three sizes: the No. 1 (a 75-pound hammer), the No. 2 (a 50-pound hammer and the number 3 (a 30-pound hammer). They were priced at $550, $450 and $350 respectively.[37] At that time, Long and Allstatter was manufacturing nearly a hundred different sizes and designs of power punch, including double machines which conserved valuable shop space. While double power hammers were built by other companies, Long and Allstatter evidently never built their helves in this configuration).[38] The manufacture of these large metal-working machines required sophisticated manufacturing techniques. "The best brands of cold blast and other iron which secures the greatest strength," was used for the heaviest castings and admixtures of steel lent additional strength and durability to such vital parts as clutches, pinions and slides. Evidently, there was some indecision as to which sizes of hammer would be most desirable to manufacturers, for the earliest selection was soon replaced by a line of four sizes: 80-, 60-, 40- and 25-pounds (priced at $700, $575, $450 and $350 respectively). This uncertainty not withstanding, the company soon boasted that "the fact they are to be found in large numbers in the most prosperous and successful shops is the best evidence of their utility and the degree of perfection to which they have been brought."[39] By 1884, Long and Allstatter had eliminated the 25-pound model, adding a 100-pound, a 30-pound, and a 20-pound hammer to complete its line. These hammers were comparable in price to the Bradley Rubber Cushioned Helves.[40] The redesigned American Rubber Cushioned Helve Hammer was built in five sizes: 100-, 80-, 60-, 40- and 25-pounds. The 100-pound hammer was said to be suited for work averaging two and one half inches square, providing a maximum of 250 blows per minute. The 25-pound hammer was suited for work one inch square, offering up to 400 blows per minute.[41]

A 60-pound model of the American Rubber Cushioned Helve Hammer. (*American Machinist*)

The American Rubber Cushioned Helve Hammer appeared in the pages of *American Machinist* in the Fall of 1920. Particular attention was drawn to the fact that "all connections and parts that have a tendency to work loose are carefully protected against such trouble" (perhaps a drawback with the previous Long and Allstatter hammers, but more likely a general complaint with this type of machine). Attention was also directed to the treadle, which was

A rear view of Long and Allstatter's American Rubber Cushioned Helve Hammer clearly shows the re-designed means of raising and lowering the helve. (*American Machinist*)

designed to take up lost motion between it and the belt-tightener, providing an action "sensitive and smooth...in regulating the force and rapidity of the blows." The hammer always came to a stop in the "up" position and used a steel brake-band with a friction lining which nearly encircled the brake wheel (opposite the main drive pulley on the drive shaft).[42] The American hammer was substantially more robust than its predecessor. The vertically-adjustable frame was now supported by a single casting which formed an arch over the moving parts. The yoke was redesigned, with large rubber cushions below the helve on either side of the pivot. A smaller rubber cushion was attached to the yoke above the helve behind the pivot, and a similar cushion was suspended from the frame in the front. The operating instructions for the American hammer give one a sense for how these cushions were adjusted and employed:

"...Lower the two bottom rubbers, so that they are free of the helve, force the upper rear rubber down on the helve until the helve hangs free with no other means of support than its pivot and this upper rear rubber. Raise the two lower rubbers against the helve, adjusting both alike, squeezing them fairly tight against the helve. The front upper rubber should be set to suit the stroke, that is, the rubber should be set so that it catches the helve just before the highest point of the stroke is reached, thereby serving as a resilient bumper and recoil, helping to return the helve on its return movement. Different work, and different speeds and strokes, can advantageously use different pressures of the rubbers, and with a little experience the operator will understand the functions of the respective rubbers so that a correct adjustment can be quickly made."[43]

The Long and Allstatter Company continued to be listed under "Hammers: Helve, Power" in the *Thomas' Register of American Manufacturers* through 1939.[44]

* * *

Charles M. Brown of Chicago, Illinois, contributed to the design of Bradley Hammers in the 1890's (a non-vertically-configured design which will be discussed in Chapter 5). But in the previous decade, he developed an unusual crank-actuated helve hammer which, for a brief period, gave the Bradley Rubber Cushioned Helve a run for its money. The Brown Slow Raising, Quick Striking Hammer was built by the Knowlton Manufacturing Company of Rockford, Illinois; and advertisements for the machine often appeared in *American Machinist* side-by-side with those for Bradley's hammer. While the most notable feature of Brown's hammer was its slow raising, quick striking mechanism, it too employed rubber cushions to isolate the machine from impact. Like the Long and Allstatter hammer, it was possible to easily raise and lower the helve to accommodate dies and work of varying thicknesses. The Brown hammer was both versatile and mechanically elegant.

The slow-raising, quick-striking feature was accomplished through a mechanism which connected the crank to the pitman. Three-fifths of each rotation of the crank was employed in raising

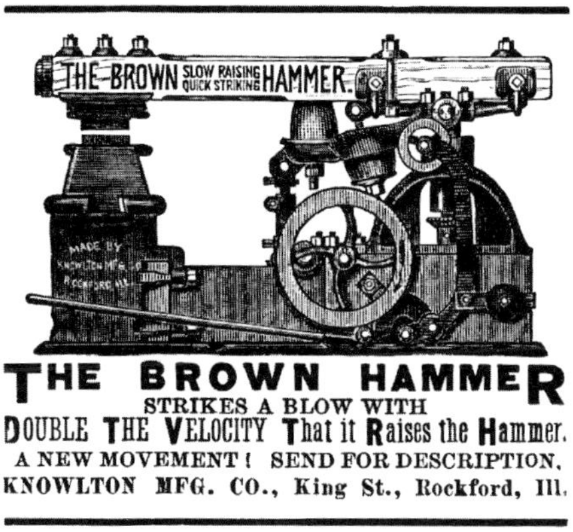

A Brown Hammer advertisement from 1885. (*American Machinist*)

the helve, while two-fifths was dedicated to drawing it down. A diagram of this ingenious mechanism was included with a profile of the hammer in *American Machinist* in the Summer of 1884: the numbers one through twenty show the manner in which rotation of the crank was translated to movement of the linkage in its arc (the threaded-rod at the left is the lower end of the pitman, adjustable to vary the force of the blows).[45]

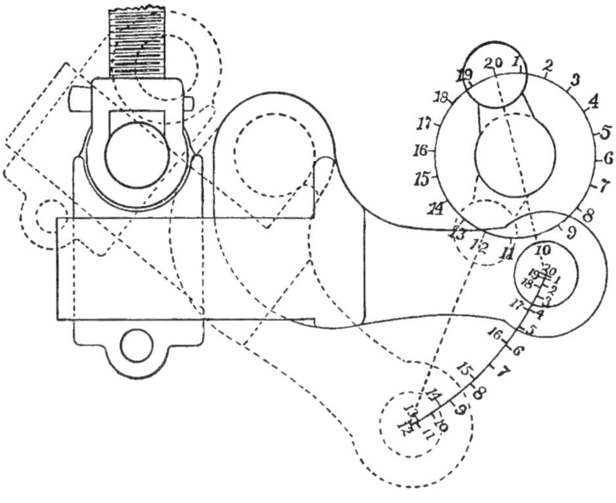

The observation that a hammer should be raised and brought-down at different rates was astute. Trip-actuated helves, which were still in use but largely obsolete, could provide such a stroke; but gravity limited the rate of descent. Some trip-actuated helves, such as those which operated at Saugus in Massachusetts during the mid-seventeenth century, included wood springs to accelerate their descent. However, those were heavy, slow hammers. The Brown hammer, like its contemporaries, was a far lighter machine, operating at a rate much faster than gravity could provide. Vertically-configured, crank-actuated hammers employ spring linkages to effectively proportion each rotation of the crank (for example, the ram continues to rise as the crank pin begins its descent, with the spring absorbing and storing energy). Early rubber cushioned helves were another matter, and more subject to the limitation of a crank moving at a constant rate. Since "the weight of helve and hammer (was) to be raised at each blow, it necessarily followed that nearly all of the power was exerted in raising the hammer and little to delivering the blow."[46]

Charles M. Brown was granted his first power hammer patent on August 3, 1880. Patent number 230,612 describes a guided-ram hammer with a two-part pivoting beam. It was cushioned by a novel arrangement of rubber springs attached to struts above and below the helve. All of Brown's designs feature an efficient use of rubber cushions, and it was such an arrangement that he later assigned to Bradley (as a possible means to tension the leather straps of the Bradley Upright Strap Hammer). Another distinctive feature of Brown's first patent was a variable pitman which could be adjusted while the hammer was running by means of a conveniently-placed hand lever. This first design also included the slow-raising, quick-striking mechanism.

A second patent was granted May 3, 1881. Patent number 240,881 describes a crank-actuated helve with numerous refinements over the previous design. The trunions holding the helve could now be raised or lowered: bevel gears linked screws on each side of the hammer to a handwheel at the center, rear of the machine. This differed from the Long and Allstatter arrangement in that the drive shaft and crank were fixed. The pivots at each end of the pitman accommodated changes in the height of the helve.

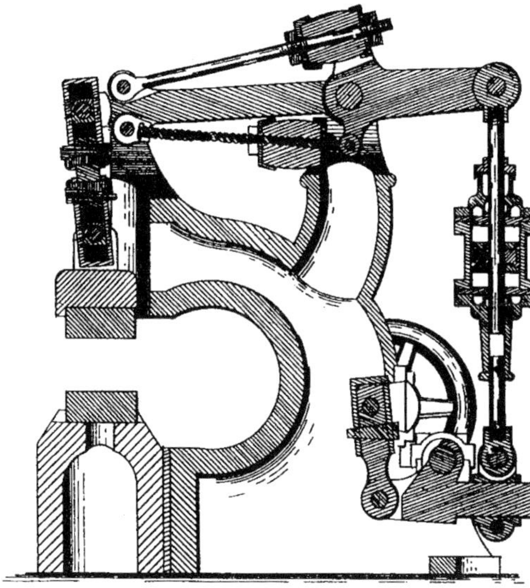

Charles M. Brown's first power hammer patent (1880) described guided-ram machine, which used a two-part beam. This was similar to Alfred Grainger's thinking back in 1867. Not shown in this cross-section is the hand-wheel-operated adjustable pitman.

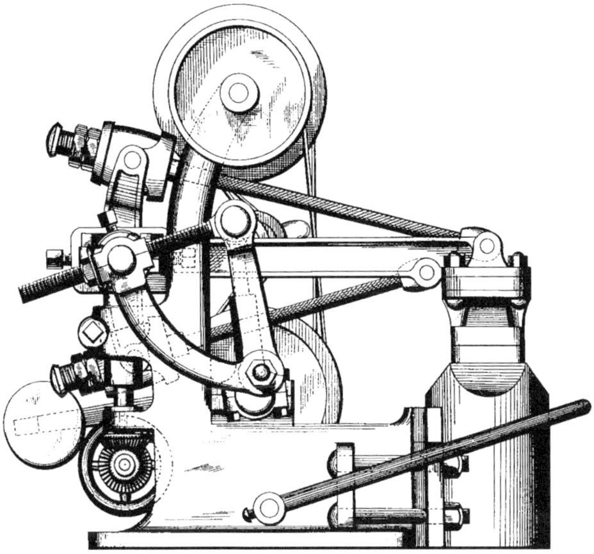

Charles M. Brown's second patent (1881) describes a true helve, one which could be readily raised or lowered to suit varying thicknesses.

The slow-raising, quick-striking mechanism was adjustable, so that it was possible to precisely vary the proportion of crank rotation dedicated to raising and striking. This was undoubtedly an important step in developing the hammer that was ultimately produced in Rockford. While the drawings that accompany Brown's early patents suggest that working models were built, it is unknown if any hammers were produced commercially prior to the Knowlton machine.

When he applied for a third hammer patent in 1885, Brown was a resident of Rockford. Patent number 327,999 was granted October 13, 1885, and described a helve very similar to the Knowlton hammer, which was already on the market. As noted, the proportion of raising to striking was fixed at 3 to 2. Unlike the production model, Brown's third design included a tilting and rotating anvil (along with a means of sliding the helve slightly forward and back to accommodate changes due to the angle of the anvil face). This made it possible to accommodate various thicknesses of dies and work without resorting to a vertically-adjustable helve. While this arrangement might have been less expensive to produce, it was probably far more cumbersome to use than the vertically-adjustable trunions, and the Knowlton

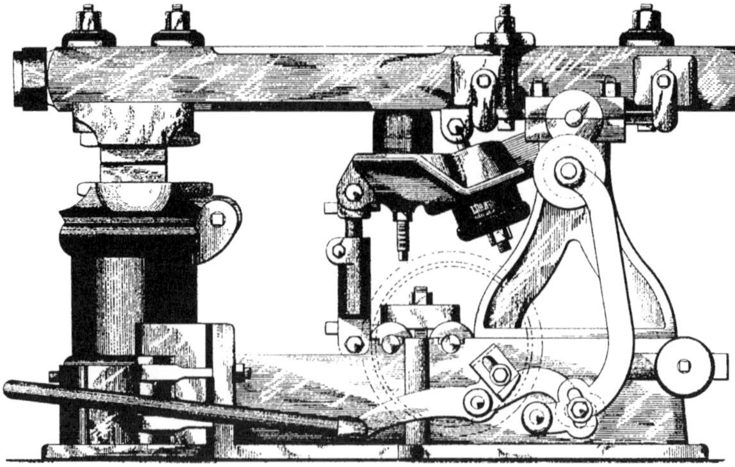

Charles M. Brown's patent of 1885 described the hammer which was to bear his name. The height of the helve was fixed, but the anvil could be tilted to accomodate (imperfectly) varying thicknesses.

Company wisely adopted the latter. The Brown Slow Raising, Quick Striking Hammer was produced through at least the Fall of 1887,[47] having been adopted for use in some of America's most important shops; among them, the Cambria Iron Works of Johnstown, Pennsylvania.[48]

* * *

Charles M. Brown was not the only inventor of this period to consider the advantages of varying the proportion of crank rotation dedicated to raising a helve hammer and striking the blow. Columbus Washington Corr of Carlinville, Illinois, was granted patent number 206,665 on August 6, 1878. While Corr's design was described at length in the pages of *American Machinist* in the Fall of 1880,[49] it is not known if hammers of this type were actually manufactured (none were ever advertised in that periodical). However, references to dimensions of parts and the weight of the anvil (500 pounds) strongly suggest that some were built. It is also not known if Corr's efforts in any way inspired those of Brown.

The helve of Corr's hammer was suspended in an "oscillator," between flat leaf springs above and below. The eccentric (crank) moved in a loose box which was adjustable, running in a "carriage" which could be shifted forward and back. This arrangement made it possible to vary the hammer's stroke to accommodate work of various thicknesses. For short strokes on thin work, the carriage was shifted back, so that the shaft of the

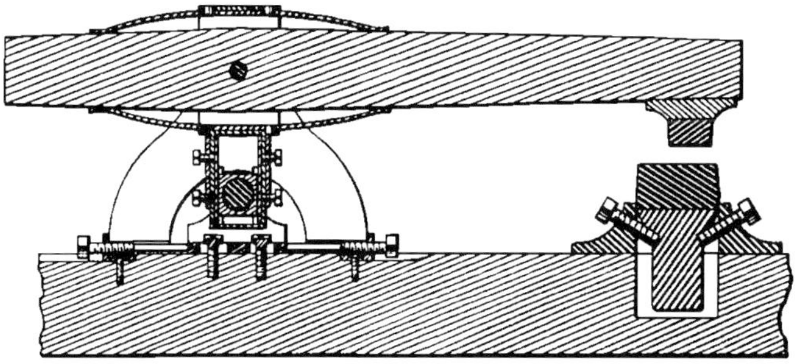

Columbus W. Corr's patent of 1878 described another helve hammer which provided variable raising and striking rates.

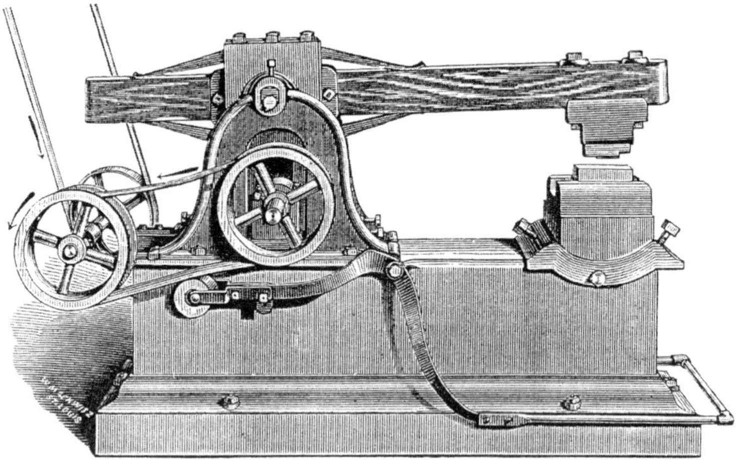

Corr's Helve Hammer. (*American Machinist*)

eccentric was behind that of the oscillator. For thick work, the carriage was moved forward so that the shaft bearing the eccentric was forward of the oscillator's pivot point. These adjustments were made with sets of bolts on either side of the carriage. This was a cumbersome arrangement, and suggests that the hammer was intended primarily for manufacturing. Corr's design yielded a downward stroke that was one sixth quicker than the up stroke: the down stroke was performed while the eccentric revolved above the center of its shaft and nearest the fulcrum of the oscillator (five-twelfths of a revolution). The up stroke was performed while the eccentric revolved below the center of its shaft, in that part farthest from the fulcrum of the oscillator (seven-twelfths of a revolution). In order to accommodate changes in the thickness of the work or dies, Corr resorted to the cumbersome device of a tilting anvil. While the oscillator and drive of Corr's hammer were promising developments, the overall machine was probably not successful.

The inventor's choice of flat steel springs to cushion his hammer is notable. In his patent, Corr claimed them to be advantageous and "unlike the rubber springs, which are of ever-varying quality, and which require frequent adjustment."[50] These shortcomings were doubtless real, but rubber springs were also extremely effective and remained popular for many decades.

*   *   *

One of the most curious American-built crank-actuated helve hammers was Penney's Spring Cushioned Helve Hammer. It employed a pivoting beam, above the helve, which imparted motion to the helve through a bow-spring linkage attached at the head. The use of this spring linkage dealt effectively with the problem of raising the helve and striking a blow with a crank moving at a constant rate (borrowing the idea from vertically-configured hammers). Another particularly interesting and important feature of Penney's hammer was the capability of moving the fulcrum of the pivoting beam to adjust the strength of its stroke. This unusual feature, and its British precedent, is discussed in greater detail in the section devoted to Arthur Alfred Koch's

Isaac L. Penney's Spring Cushioned Helve Hammer used a beam above the helve to impart motion from the crank. Furthermore the fulcrum of the beam could be adjusted to vary the strength of the blow. (*American Machinist*)

movable-fulcrum hammer (a non-vertically-configured, guided-ram machine) in Chapter Five. While the duration of the manufacture of Penney's hammer is not known, it was built in at least four sizes, and it was offered for a time by the respected New York machinery dealer Manning, Maxwell & Moore.

The salient features of Penney's hammer were described in U. S. Patent number 302,520 which was granted to Isaac L. Penney of Minneapolis, Minnesota, on July 22, 1884. The patent application was filed in August of 1883, at which time the hammer was already in production. A brief notice had appeared in the "business specials" column of *American Machinist* in July of that year, placed by H. E. Penney, "manufacturer of Spring Cushioned Helve Hammers" in Minneapolis.[51] The rights to build the hammer were soon sold, for in March of 1884 the hammer was offered by O. L. Packard of Milwaukee, Wisconsin. Advertisements identified O. L. Packard as builders of iron and woodworking machinery, and as a source for engines, boilers and supplies.[52] This company was definitely familiar with power forging hammers, having been listed among early users of the Shaw and Justice Dead Stroke Power Hammer in the 1870's.[53] The Penney hammer was briefly profiled in *American Machinist*, which noted that it is "intended for hard work and rough usage." In addition to the variable stroke, attention was drawn to the fact that the husk (pivots) could be raised or

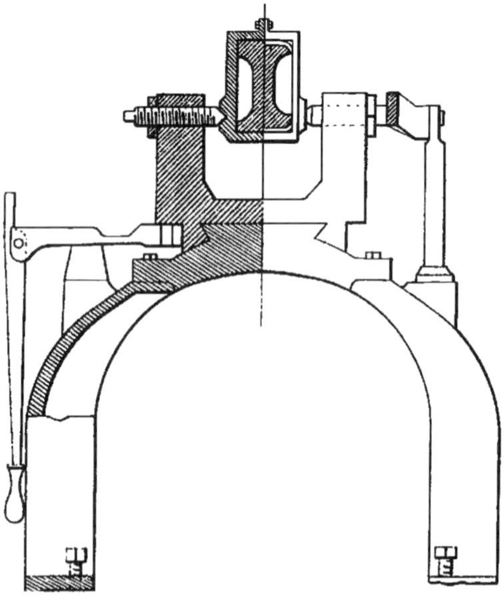

A detail from Isaac Penney's patent of 1884 shows a cross-section of the movable fulcrum. The drop-away lever at the left was brought up to move the fulcrum along a dove-tail slide.

lowered to accommodate various thicknesses of work and dies. However, the blacksmith had to adjust each side separately, which must have proven cumbersome. The speed of the hammer was regulated by a "paper friction" which engaged as the foot treadle was depressed.[54]

The Manning, Maxwell & Moore catalogue of September, 1884, reveals that there were four sizes of the Penney hammer available: in 40-, 60-, 80- and 100-pound sizes. The largest stood five feet eight inches tall, weighed 5,000 pounds overall, provided 200 blows per minute and boasted a capacity of five and one half inches for wrought iron (five inches for steel). The 100-pound hammer cost $625. The smallest model stood four feet ten inches tall, weighed one ton, provided 325 blows per minute and was designed for wrought iron stock up to two inches (one and a half inches for steel). The 40-pound hammer cost $300. On the basis of hammer head weight, the Penney hammers cost only about two-thirds the amount of the Bradley and Long and Allstatter machines.[55] Despite this advantage, and the machine's unique features, advertisements for the hammer soon disappeared.

* * *

Relatively few power forging hammers were manufactured west of the Rocky Mountains, but one exception was the Holt Power Hammer. Built by the Holt Manufacturing Company of Stockton, California, circa 1905-1915, this machine used rubber cushions in a manner reminiscent of the early Butterfield and Hay patent. The Holt crank-actuated helve was extremely robust. It was built in one 75-pound size, and weighed 3,640 pounds overall. The machine tool department of the Holt Manufacturing Company claimed that the hammer was "the outgrowth or result of necessity. Being unable to find a power hammer that would stand the strain of continuous work under all conditions, we were forced to design and build one."[56] It would be interesting to learn which hammers had failed to satisfy the Holt engineers. One clue may be gleaned from the description of the rubber cushions, or "plungers,"

The Holt helve hammer probably played a role in the development of the caterpillar tractor and its military extrapolation: the tank. (*Holt Company catalogue*)

which were said to eliminate the "annoyance due to the breaking of springs."[57] The Holt helve hammer was touted for its "great adaptability for all classes of forgings; from light die work to heavy agricultural forgings." To accomplish this, the standards were adjustable so that, upon loosening three bolts, the helve could be easily raised or lowered. The capacity of the hammer was rated up to four inch round iron, with a stroke from one to seven inches.[58] There was apparently no patent associated with this machine, despite the claim that "new and important improvements" had been added in its development.

The Holt Manufacturing Company is legendary in the history of American agricultural machinery. In 1925 it merged with the C. L. Best Tractor Company to become the world-renowned Caterpillar Tractor Company. During the "Big Steam Era," 1890-1910, Holt and Best built combine harvesters on the Pacific Coast which were capable of covering 125 acres in a day, harvesting 4,000 bushels of wheat! There can be no doubt that the Holt shops were truly the scene of "continuous work under all conditions." In fact, shortly before the hammer was placed upon the market, a momentous development occured at Holt: on November 24, 1904, the first practical track-type tractor was tested.[59] The "caterpillar" design was originally developed to solve a local problem. The rich but spongy peat soil under cultivation just west of Stockton could not support the weight of a conventional tractor. The solution to this agricultural problem opened-up a realm of almost limitless potential. In 1915, the Holt Manufacturing Company developed a radical new military vehicle. Production of the so-called "land-destroyer" began the next year. For secrecy's sake, when first shipped, they were labeled as 'water tanks' for the British troops in Egypt. The name 'tanks' stuck. Prior to this vital defense production, the Holt Manufacturing Company took its power hammer seriously. The product was listed in the *Thomas' Register of American Manufacturers* in its fourth (1909) through sixth (1914) editions.[60] It is unlikely that any hammers were built after 1915.

In addition to the Holt Power Hammer, a surprisingly large number of new crank-actuated helve hammers appeared in the early twentieth century. There were still many trip-actuated helves to be seen pounding away in shops scattered throughout the United States, and sales of the expensive, high-quality Bradley helve hammers remained strong through the 1920's. "It Looks Like a Hammer," ran the advertising slogan for one turn-of-the-century helve (the Dayton Power Hammer); and, indeed, many American blacksmiths put their trust in the familiar form. Some crank-actuated helve hammers were also among the most affordable hammers for the small-scale blacksmith. Many American smiths had just installed gasoline engines in their shops and were looking for a first power hammer.

While the Holt hammer used rubber cushions, the majority of new designs used metal springs to generate a snappy blow and to protect the machine from its inherent self-destructive tendencies. This was accomplished with various spring linkages, many of them derived from those first used in vertically-configured guided-ram hammers. In the pages which follow, crank-actuated helves using bow-shaped, flat springs will be considered together first, followed by hammers which use coil springs. These relatively straightforward helves were succeeded by more-sophisticated "mechanical strikers" and compound helves (helves capable of maintaining parallel dies throughout the hammer's stroke).

* * *

Hathorn's Hard Hitting Helve Hammer was the invention of Henry W. Hathorn of Mason City, Iowa. Granted U.S. patent 758,361 on April 26, 1904, the design was both simple and clever. It incorporated a means to compensate for various thicknesses of work, as well as the ability to adjust the force of the blow while the machine was running. The hammer was a success from the start. It was manufactured by the Hathorn Foundry and Machine Company until 1908, when the rights were sold to the West Tire Setter Co., of Rochester, New York. The hammer then became the Rochester Helve Hammer; in which guise it was modified and manufactured through the late 1930's.

Hathorn arrived in Mason City in a covered wagon in 1877. Purchasing "a piece of wild land" in Cerro Gordo County for five dollars an acre, he set to farming. He moved into Mason City in

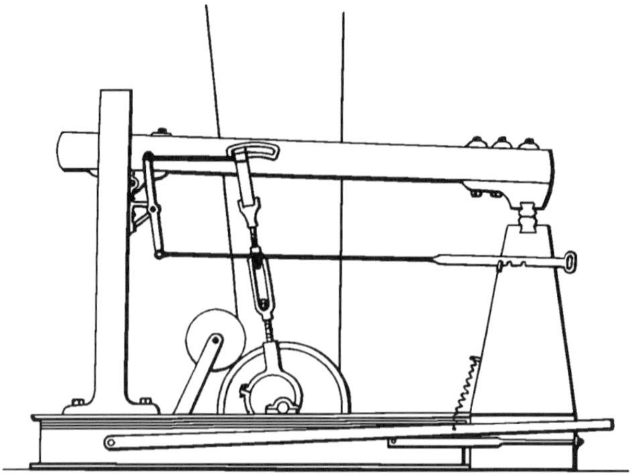

Henry Hathorn was granted a patent on his Hard Hitting Helve Hammer in 1904.

1890, and entered the foundry business.[61] In 1904, he sold his foundry in order to concentrate on the manufacture of his hammer. The purchaser was none other than Arthur A. Koch, inventor of the Modern Power Hammer, and the old foundry became the Vulcan Iron Works. It was reported that Hathorn moved to Grinnell to pursue his hammer-building, and this is confirmed by advertisements in *The American Blacksmith*.[62] However, subsequent advertisements in *The Blacksmith and Wheelwright* continue to refer to the Hathorn Foundry and Machine Company of Mason City. Additionally, the *Thomas' Register of American Manufacturers* places the Hathorn Foundry & Machine Company in Chicago in its third (1907-1908) through fifth (1912) editions. It is quite possible that Hathorn's hammers were manufactured in Mason City (for Hathorn) by Koch, but this has yet to be confirmed.

Hathorn's hammer was built in two sizes: size "A" weighing 1,000 pounds over-all, and size "B" weighing 1,350 pounds (ram weights were not specified). Advertisements in *The Blacksmith & Wheelwright* boasted that the hammer could weld tires up to eight

Hathorn's Hard Hitting Helve Hammer. (*The Blacksmith & Wheelwright*)

inches wide and 1 1/2 inches thick, though it was probably best suited to work of a more modest scale. A simple turn-buckle allowed the rod which actuated the helve to be lengthened or shortened to suit the thickness of the work. A lever, conveniently mounted on the side of the anvil, adjusted the force of the blow. This was accomplished by shifting the point at which the actuating linkage connected to the helve: the closer to the pivot, the stronger the blow.[63] With the sale of the rights to his hammer, Hathorn entered the employ of the West Tire Setter Company,[64] but his association with the firm was probably brief. He subsequently opened a machine shop in Davenport, Iowa, but returned to Mason City in 1921. It was said that Henry Hathorn repaired the first automobile that came to Mason City: a one-cylinder Cadillac, and that one of his sons, Charles, started the first garage there.[65]

In the hands of the West Tire Setter Company, the helve hammer line was immediately expanded. The size "A" hammer, with a 25-pound head, could handle stock up to one and three-quarters of an inch square, and cost $200. It came with drawing dies, but an extra set of welding dies could be had for $10. The upper die was not to exceed eight pounds; but if heavier dies were necessary, an extra-heavy-duty spring could be installed at the factory. The largest regular model, the size "F," offered a 100-pound head, and could handle stock up to four inches square. It cost $850, and it was recommended that the weight of the upper die not exceed 30 pounds. Between these models were hammers with heads weighing 35, 50, 60, and 80 pounds, priced at $250, $500, $600 and $700 respectively. The smallest hammer ran at a maximum speed of 400 blows per minute, which diminished to 250 for the size "F." In an earnest attempt to enter the industrial helve hammer market, and compete with the Bradley Company, the West Tire Setter Company created a new line of so-called "Manufacturer's Type" hammers. Built in one basic size (weighing from 4,600 to 4,600 pounds) the hammer could be purchased with a head weighing 60, 80 or 100 pounds, priced at $650, $750 and $900. This hammer boasted a substantial cast iron frame which was reminiscent of that used in the Bradley Rubber Cushioned Helve. The company was

The "Manufacturer's type" of the Rochester Helve Hammer. (*The Blacksmith & Wheelwright*)

pleased to point out that there were "no rubber cushions to be replaced from time to time." Unlike the lighter-duty Rochester Helve Hammer line, the Manufacturer's Type offered a vertical adjustment of the pivot (five inches), but it was not possible to adjust the stroke of the hammer "on the fly." Illustrations in a catalogue of 1911 reveal that the lighter Rochester Helve Hammers could also be had in a non-shiftable configuration, with the spring connection bolted to the helve, instead of readily-movable as in the original Hathorn design.[66] However, the hammer was evidently overpriced for the general blacksmithing market. In November of 1912, the company took out a full page ad. in *The Blacksmith & Wheelwright* offering the size "A" hammer for $130. It was claimed that this 25-pound hammer could do more and heavier work than some of the 50-pound hammers then on the market: weld 2-inch axles, weld tires up to four inches wide by one inch thick "and do a good job."[67]

In its move from the Midwest to the East, a transformation occurred. The West Tire Setter Company touted the Hathorn patent as "superior in design and mechanical principles," but claimed to have contributed "a thoroughness and accuracy of workmanship and an excellence of materials."[68] The hammer had indeed found a good home. The West Tire Setter Company had been founded half a century before by an inventive genius, Jonathan Burns West. Among his earliest inventions were a machine for turning broom handles, a water meter, and an embroidery machine. His most famous and successful machines were designed for the rapid assembly of wagon and carriage tires. The first such device, West's American Tire Setter, achieved worldwide acclaim at about the time Bradley's Rubber Cushioned Helve Hammer was introduced. About 1880, a steam-cylinder-actuated model was introduced, to be ultimately superceded by a hydraulic model.[69] J. B. West died in 1900, long before the company acquired Hathorn's hammer. Had he been born a generation later, his name would probably have appeared in the hammer-building patent record. The West Tire Setter Company was listed under "hammers, power" in the *Thomas' Register of American Manufacturers* through its 27th edition (1937).[70]

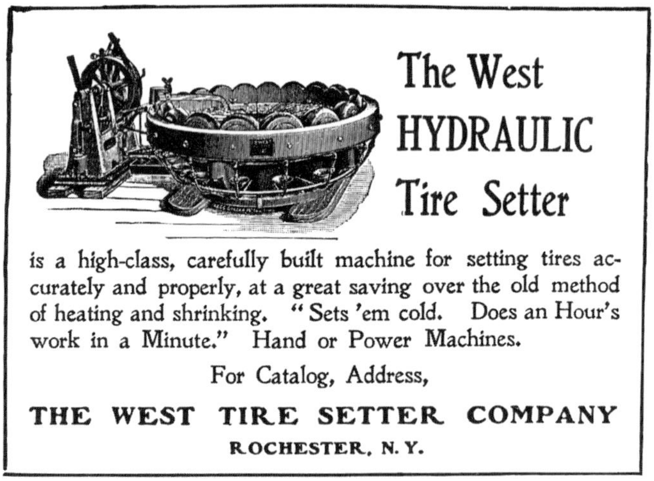

An advertisement for an improved model of West's Tire Setter. (*The Blacksmith & Wheelwright*)

* * *

Bicknell's Power Hammer was introduced in 1903 with flat springs connecting its crank to the helve. It was claimed that this machine had fewer parts than any other hammer.[71] By 1909, the steel springs had been replaced by rubber cushions, and the machine had become Bicknell's Triple Helve Hammer (a reference to the cushioning arms extending above and below the helve).[72] The hammer remained in production through at least 1918.[73]

An early (1903) advertisement for the Bicknell Power Hammer. Note the flat steel springs above and below the helve. (*The Blacksmith & Wheelwright*)

The Company was founded in Janesville, Wisconsin, about 1889, as Bicknell Brothers - Wagonmakers and Supplies, by Henry S. and Raymond C. Bicknell. It soon became the Bicknell Hardware Company; and, ultimately, the Bicknell Manufacturing and Supply Company. In addition to the helve hammer (which was probably offered in one size only), the company manufactured a line of woodworking machinery for carriage-makers. Included was a combination jointer which could be fitted with attachments for tennoning and boring, and felloe-rounding. Another machine combined table saw, jointer, felloe-rounder, boring and tennoning machine and tire-bolter. Alongside this multi-purpose wonder, the helve was described as "simple, effective, practical."[74] The company diversified into automotive products, and even manufactured spark

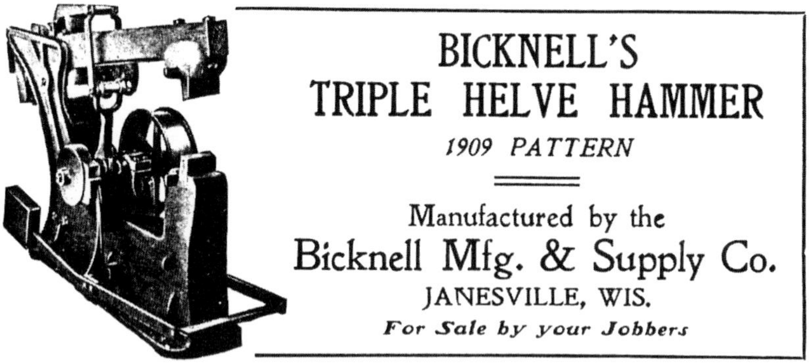

A Bicknell advertisement from 1909. By this time the hammer was cushioned with rubber springs. (*The Blacksmith & Wheelwright*)

plugs.[75] The company prospered in the twenties, and opened a branch in Beloit, Wisconsin. The Bicknell Manufacturing and Supply Company weathered the Great Depression, and it was last listed in the Janesville City Directory in 1939 (as an automotive accessories store).[76] It appears that there was no patent associated with this hammer.

* * *

A number of the crank-actuated helves which employed flat springs were developed to complement previously existing hammer lines. The following three machines were built by companies which were better known for their guided-ram hammers. For that reason, these companies will be more thoroughly covered in the chapters which follow.

Heinrich Feldhus of Hallam, Nebraska, started out in the power hammer business after submitting a drawing of his own homemade hammer to *The Blacksmith and Wheelwright.*[77] The idea of employing "I" beams for the frame of a vertically-configured hammer was so well-received that he filed for a patent (number 716,019 which was granted December 16, 1902). This hammer was soon on the market as the Little John Power Hammer, and Feldhus quickly applied the same economical construction technique to a crank-actuated helve. The Blacksmith's Friend Power Hammer

The Blacksmith's Friend used the same basic I-beam construction as the Little John Hammer. (*The Blacksmith & Wheelwright*)

appeared in the pages of *The Blacksmith and Wheelwright* in August of 1903. It used "I" beams for both the bed-plate and the standard which supported the pivot of the helve. The hammer used a bow-shaped spring linkage on the pitman. Like the Little John hammer, the Blacksmith's Friend used a friction clutch, which was foot-controlled, to operate the hammer.[78] Manufacture of the Little John Power Hammer soon moved from Hallam to Lincoln. The fate of the Blacksmith's Friend is unknown; though, in the absence of subsequent advertisements (and given the fierce nature of competition at this time), it was probably short-lived.

The Keller Brothers of Bloomer, Wisconsin, built a crank-actuated helve which they dubbed "Thor." Of awkward proportions, it appeared briefly, in 1904, in advertisements for the Bloomer Machine Works' Comet, a dramatically-styled non-vertically-configured, guided-ram hammer which was manufactured for more than a decade.[79] The Thor used an idler pulley to tighten the drive belt and control the hammer. This was a curious feature given the fact that the Comet employed a patented friction clutch. Like Feldhus' Blacksmith's Friend, the Thor hammer used a bow-shaped spring linkage; but, in this case, it was attached near the end of the helve (with the pivot near the center). It is unlikely that the Thor helve hammer was on the market very long.

Thor! (*The Blacksmith & Wheelwright*)

George E. Davis of Dubuque, Iowa, had already achieved considerable success with his vertically-configured Boss Power Hammer when he introduced the Helva-Hard-Hitter. It was placed on the market in the Spring of 1908;[80] and like the Boss, it was protected by patent. The hammer was configured much like the Keller Brothers' Thor, with a bow-shaped spring linkage attached to the end of the helve. Unlike either the Thor or the Blacksmith's Friend, the Helva-Hard-Hitter had no unifying bed-plate. Patent number 888,391 which was granted on May 19, 1908, described Davis' unusual helve. The helve itself was composed of two parallel steel bars which were held apart at intervals by spacers. This

The Helva-Hard-Hitter. While not a "watch charm variety" hammer, it was among the smallest built. (*The Blacksmith & Wheelwright*)

arrangement facilitated the use of a double-ended head which could be easily reversed for either general forging or plow work. The hammer was controlled by a foot-operated idler pulley, and it could be adjusted for varying thicknesses by means of a telescoping pitman. Despite an outrageous claim that the Helva-Hard-Hitter could forge material five inches thick, Davis realistically observed that "many shops do not need a hammer for extra heavy work, and some shops do not have steady work for a power hammer, and it is this class of trade that will welcome the opportunity to buy the Helva-Hard-Hitter."[81] Though the hammer was said to be far less expensive than the Boss power hammer, the exact cost is not specified in its advertisements. The Helva-Hard-Hitter was evidently quite successful. It was built by Davis at the Novelty Iron Works in Dubuque for several years, and remained in production when the rights were transferred to the Armstrong Manufacturing Company of Waterloo, Iowa. When the Boss hammer moved a second time, to the Novelty Iron Works of Dyersville, Iowa, the Helva-Hard-Hitter again followed. In April of 1917, a Boss number

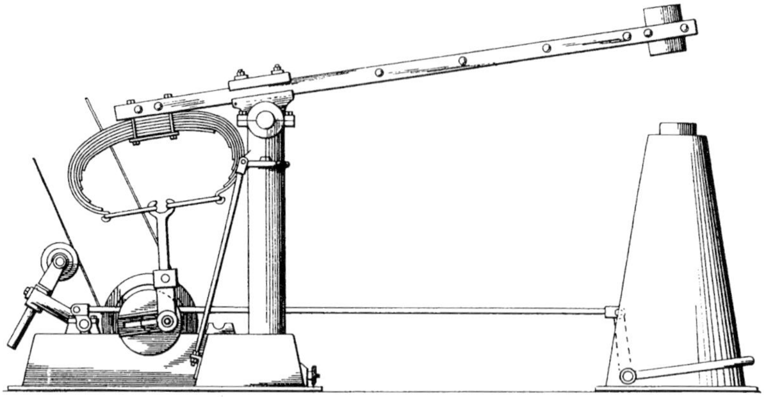

The Helva-Hard-Hitter was patented by George E. Davis in 1908.

one hammer (with a 25-pound head) cost $85.[82] The Helva-Hard-Hitter must have been cheap indeed! After 1917, advertisements for the Boss and Helva-Hard-Hitter suddenly disappeared from the pages of *The Blacksmith and Wheelwright*, and the fate of the little helve hammer is unknown. In 1923, an isolated advertisement for the Boss hammer appeared (built by the Eureka Manufacturing Company of Belmond, Iowa), but there is no mention of a helve.[83]

* * *

The crank-actuated helve built by James Depew of Loup City, Nebraska, reflected the influence of the Helva-Hard-Hitter. Depew was granted patent number 1,076,554 on October 21, 1913, for a hammer which also used parallel steel bars to compose its helve. Like Davis, Depew used this design to employ a reversible hammer head. The novelty of Depew's invention lay in its friction clutch, the interesting manner in which the bow-shaped spring was connected to the pitman, and certain details of assembly. For instance, the inventor drew special attention to the manner in which the pivot was supported and attached to the frame. This design detail was both economical to manufacture and easy to assemble. In 1914, a notice appeared in *The Blacksmith and Wheelwright* describing the products of the J. I. Depew Disc

Sharpener Company. A "trip hammer" is mentioned but not illustrated.[84] As the company name implied, Depew's most important product was a sharpening machine for plow and harrow discs. The disc sharpener described in 1914 was versatile, allowing the edge be be sharpened by either rolling or cutting (depending upon the type and condition of the disc). The machine was claimed to be "durable, compact, sold at a reasonable price... a child can operate it." At this point, Depew was in direct competition with the hammer-sharpening fixture built by MacGowan and Finigan for their Perfect Power Hammer. Other disc-sharpeners similar to Depew's were manufactured by hammer-building concerns such as the Mayer Brothers and the Star Foundry and Machine Company. Apparently the Depew Disc Sharpener was one of the best. The machine survived into (at least) the early 1930's, at which time advertisements in *The Blacksmith and Wheelwright* also mentioned the "trip hammer," available then in two sizes.[85]

Depew's power hammers were mentioned in advertisements for the disc sharpener, but were not illustrated. (*The Blacksmith & Wheelwright*)

Plow and harrow discs could also be sharpened with a hammer. Interestingly, Depew was not the only hammer-builder to combine these product lines. (*The Blacksmith & Wheelwright*)

In 1924, James Depew had been granted a second power hammer patent (number 1,485,486 on March 4). This patent described a smaller hammer, essentially a bench-top model, which employed a novel means of securing the dies and a different (but

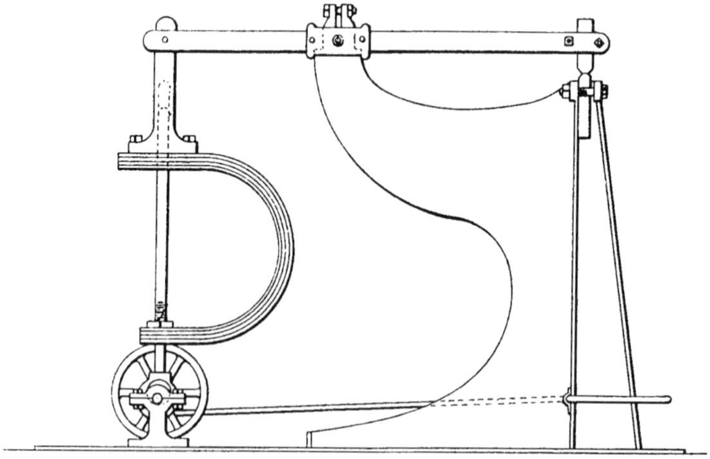

Depew's first patent (1913) described an unusual spring linkage, and a helve much like that of Davis' Helva-Hard-Hitter.

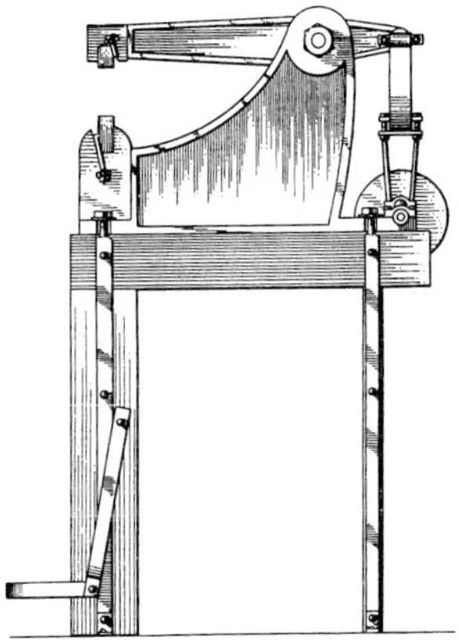

Depew's second hammer patent (1924) describes a bench-top machine.

more conventional) means of employing the bow-shaped spring cushion. Instead of securing the dies in the hammer head and anvil with tapered wedges, Depew used "U" bolts. When tightened on either side, these "U" bolts wedged the dies in place and eliminated the possibility of lateral movement. The orientation of the bow-shaped spring, with ends curving upward, is reminiscent of the earliest models of the Kerrihard and Modern power hammers. The actual appearance of Depew's helve hammers is unknown to the author, as is their ultimate fate.

* * *

In Mankato, Minnesota, the Mayer Brothers' model "E," or Easy Helve Hammer, used a linkage with a single coil-spring in lieu of a bow-shaped spring. Like the Helva-Hard-Hitter and the Blacksmith's Friend, the Easy was developed to complement an existing vertically-configured guided-ram hammer. The Mayer Brothers' Little Giant Power Hammer had been in production for seven years when, in September of 1902, the helve was introduced.[86] This move was clearly an attempt to broaden the company's share of a rapidly-expanding power hammer market, and it reflects the strength of demand for the venerable helve design. The Easy Helve Hammer was successful to a point, but it offered

The Mayer Brothers' Easy Helve Hammer cost slightly more than the smallest Little Giant. (*The Blacksmith & Wheelwright*)

little lasting competition to the Little Giant. In January of 1903, with a full-page advertisement in *The Blacksmith and Wheelwright*, the Mayer Brothers' boasted of 550 Little Giants in use in the United States and Canada. By that time, 50 Easy helves had been sold.[87] A total of 395 of them were sold before manufacture of the model ceased in 1919.[88] Evidently there was a substantial inventory of these hammers in Mankato long after that date. A company price list from February, 1923, includes the Easy, at $150.[89] The popular 25-pound Little Giant was then selling for $10 less! The Easy was, in fact, 200 pounds heavier than the smallest Little Giant, and was fitted with a 35-pound ram. In the face of stiff competition, the Mayer Brothers' helve hammer was probably overpriced for its type. It was well-built and boldly "guaranteed FOREVER against defective material and workmanship,"[90] but the Easy did not please all who purchased it. F. R. Winters, of Texas, wrote that his Easy could do the work of three men, but recommended that his brother smiths purchase a heavier hammer.[91]

As one might expect, the Easy employed the same coil-spring linkage that was in use on the highly successful Little Giant. A patent had been granted to Louis Mayer in 1901 for his variation on the coil-spring-linkage theme (first developed by Dupont in 1890). Beyond this, a patent had been filed for some unspecified facet of the Easy's construction (prior to January, 1903), but apparently it was never granted. Perhaps this was due to a prior filing (May 13, 1902) for the somewhat similar helve designed by Davidson and Strite, of Tama, Iowa.

*　　*　　*

The Hawkeye Manufacturing Company of Tama, Iowa, first advertised its Improved Hawkeye Power Hammer in *The Blacksmith and Wheelwright* in January of 1902. It had evidently been in production for some time, for the company boasted that it was "now in use in nearly every state in the Union."[92] The advertisement also stated that a patent application had been filed. However, it is interesting to note that the patent which was ultimately granted to the builders of this hammer (mentioned above) was not filed until four months after this advertisement. Apparently, there had been a previous patent application; and, indeed, upon closer examination of the illustration of this early Hawkeye, there is an important difference. The spring linkage on the pitman employs long coil springs oriented vertically instead of the patented toggle-arm

This early (1902) advertisement for the Improved Hawkeye Power Hammer shows its unusual coil-spring linkage. (*The Blacksmith & Wheelwright*)

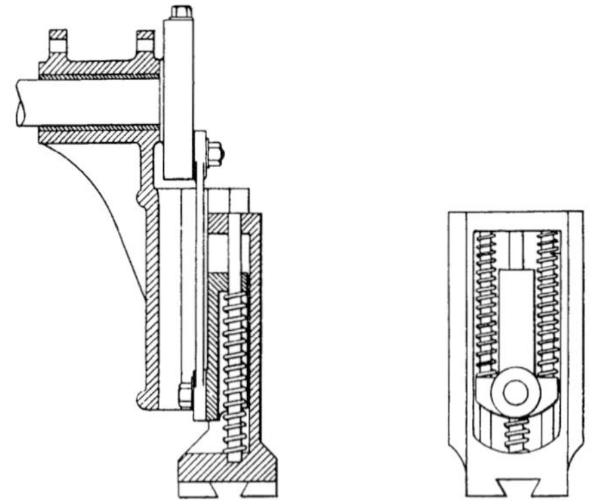

The coil-spring linkage described in Arthur Johnson' patent of 1902 may have preempted an early Hawkeye patent application.

arrangement (with two short coil springs oriented horizontally). Assuming that this linkage design was the focus of the earlier patent application, there were a number of previous patents that likely caused the application to be rejected. For example, there had been a very recent patent (number 693,324 filed April 30, 1900, by Arthur Johnson of Chicago) which featured a coil-spring linkage applied to a vertically-configured guided-ram hammer. The early Hawkeye design appears to have the same three-spring arrangement. The actual circumstances surrounding the development of this Hawkeye hammer are not known. In fact, the early history of the company was a vague subject even with those involved in the firm. For at least five years running, its advertisements boasted that the company had been in business for 12 years![93]

Frank F. Davidson and George T. Strite were granted patent number 739,992 on September 20, 1903 for the Hawkeye's toggle-arm linkage. While the stroke could not be adjusted with the hammer in operation (as with Hathorn's design), the connection point of the pitman to the helve could be moved fairly easily, and the pitman could be lengthened (for gross changes in the thickness of the work) by means of a threaded rod. The patent also depicted a simple brake which bore against the over-wide rim of the drive pulley when the foot-treadle was released (a similar

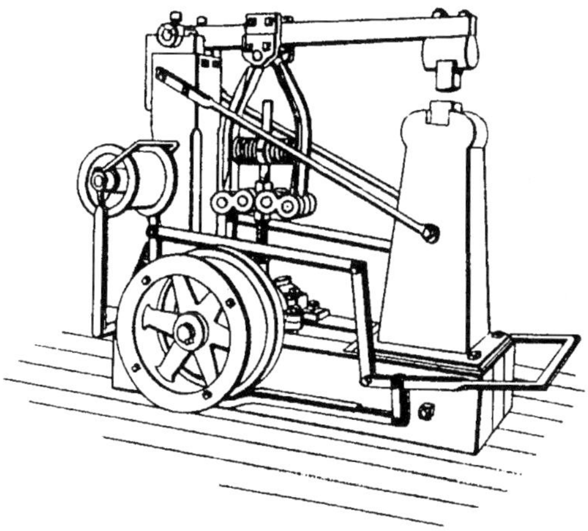

The long coil-spring linkage depicted in the earliest Hawkeye advertisements was quickly replaced. A patent for the new linkage was granted to Frank Davidson and George Strite in 1903.

arrangement was used in Koch's Modern Power Hammer, which came out shortly after the Hawkeye). Davidson and Strite's new design had appeared on the market immediately after the patent application was filed, and the word "improved" was dropped from the name. At first, only one model of the Hawkeye Power Hammer was offered, at a tempting price of $85.[94] In December of 1902, blacksmiths considering the purchase of a Hawkeye Power Hammer were urged to buy quickly, for the price would be increased to $100 in 1903.[95] Within the next year or so, the Hawkeye line was expanded. Two models were offered: the No. 2, designed for plow and general repair work, and the No. 3, designed for large shops and factories. It was claimed that the No. 3 hammer, which weighed 1,250 pounds overall and cost $150, could successfully work iron from three to four inches in diameter.[96] The No. 3 hammer was considerably more substantial than the No. 2, which weighed only 800 pounds. Interestingly, the weights of the heads alone were never specified. While both hammers were designed to sit atop a bed-plate of three stout timbers, the No. 2 hammer also used wood for the standard that supported its pivot. The standard of the No. 3 hammer was cast iron.

In November of 1906, it was announced that the Hawkeye Manufacturing Company had moved from Tama to Cedar Rapids, Iowa.[97] Soon after, the company printed a glowing testimonial for its No. 2 hammer. J. R. Allen, of Calhan, Colorado, wrote:

> "I have made and put on seven new shares a day with the hammer and sharpened five shares per hour. I weld buggy and wagon axles without a helper. The hammer paid for itself in six months. I would have paid two men $5.00 per day to do the work I did with the hammer. I can control the hammer with the greatest of ease and can crack walnuts with safety."[98]

The savings in labor costs and time that could be achieved with the use of a power hammer were significant indeed. By 1910, the company was advertising three sizes,[99] but the specifications of the new size are not known. The Hawkeye Manufacturing Company continued to advertise its machines in *The Blacksmith and Wheelwright* through the summer of 1918. Though listed in the *Thomas' Register of American Manufactures*, the company did not appear beyond the 9th, 1917, edition.[100]

A Hawkeye Power Hammer built in Tama, Iowa, prior to the move to Cedar Rapids. (*The Blacksmith & Wheelwright*)

* * *

The Dayton Power Hammer was developed by Washington Foglesong of Dayton, Ohio, just before the turn of the century. He was granted patent number 630,041 on August 1, 1899, for a coil-spring-cushioned helve hammer which must have appeared familiar to many hammermen. Fogelsong's rocking carriage design was in fact much like the hammer mechanism patented in 1883 by Alexander Beaudry (and used in Beaudry's first successful line of rubber-cushioned, non-vertically configured guided-ram power hammers). Fogelsong's later version of this design was simple, elegant, and well-suited to the relatively small hammers that employed it.

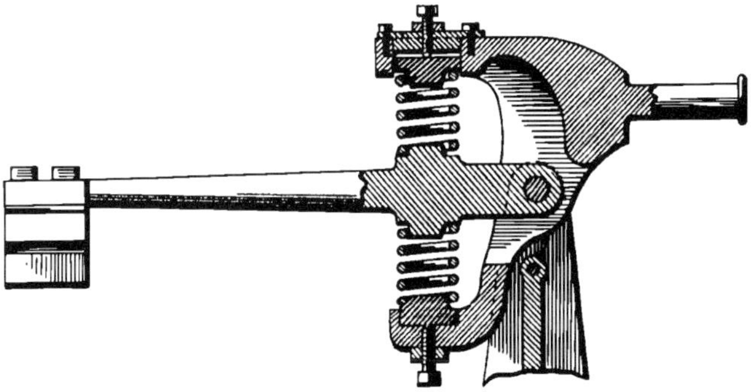

A detail from Washington Fogelsong's forging hammer patent of 1899. The design is very similar to a leather-working hammer patented by Fogelsong a decade earlier (illustrated on the next page).

Washington Fogelsong first appeared in Dayton in about 1880, as a partner in the firm of Isaac Haas and Company, manufacturers of horse collars and leather fly nets.[101] By the end of the decade, Fogelsong was in a partnership, building machinery for this specialized trade; and he was also president of the Union Collar and Net Company. In 1888, he patented a special hammering machine for making horse collars. He remained closely identified with the horse collar machinery field into the twentieth century; and his firm, the Fogelsong Machine Company, persisted beyond his involvement. The helve hammer was initially manufactured and sold by the Patterson Tool and Supply Company of Dayton. This company had been founded in 1882, and incorporated in 1897. It produced an extensive line of iron and woodworking machinery, in addition to the hammer.

A catalogue from 1901 reveals that the Patterson Tool and Supply Company was building the Dayton Power Hammer in one size, weighing 1,000 pounds overall. A relatively small hammer with a hammered steel helve, the Dayton could "readily forge 2-inch square machine steel, one foot long, at one heat."[102] In 1904, the Dayton helve hammer was offered to a wider audience. A profile appeared in *Modern Machinery*,[103] as well as *The Blacksmith and Wheelwright*. In the latter publication, the manufacturers cite one of their customers, "a horseshoer... finds that he can, at the present price of steel, economically work up old shoes into rod stock and produce a fine article."[104] By this time the hammer was being built in two sizes (weighing 900 and 1,500 pounds overall), and the manufacturer was now the Miami Valley Machine Tool Company of Dayton. This company first appeared in the Dayton directory for 1903-1904, owned by the Patterson Tool and Supply Company.[105] By the summer of 1909, the hammer was being manufactured directly by the Fogelsong Machine Company. An advertisement in the 1909 *Thomas' Register of American Manufacturers* includes the

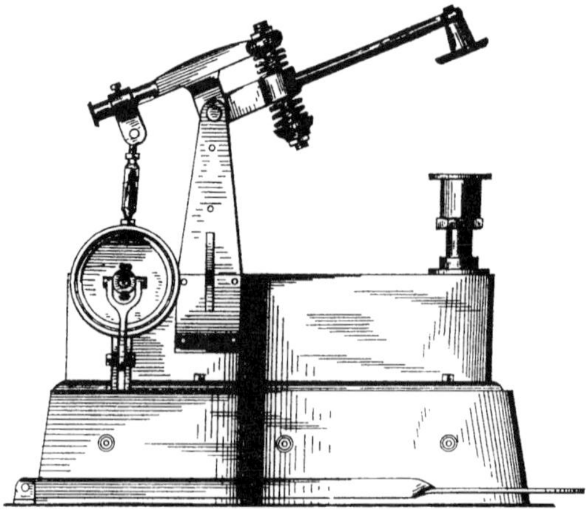

Fogelsong's hammer patent of 1888 was for a leather-working "horse collar" hammer. Its elegant coil-spring cushioning was likely influenced by Alexander Beaudry's guided-ram hammer designs.

Fogelsong's hammer was marketed as the Dayton Power Hammer. (*The Blacksmith & Wheelwright*)

perceptive Dayton slogan: "It looks Like a Hammer."[106] This company, and its spring-cushioned helve hammer, was listed through October, 1916.[107]

* * *

The Welch Variable Blow Power Hammer marked an ambitious, but short-lived, attempt to enter the heavy helve hammer market with a coil-spring cushioned design. Manufactured by the Welch Hammer and Machine Company of Cleveland, Ohio, the hammer was unveiled in the Spring of 1911, with profiles in *Machinery*[108] and the *American Machinist.*[109] A single 60-pound size was described, but the company announced that it intended to manufacture its hammer in all "standard sizes," as well as with guided-heads (like the Bradley Upright Hammers). The proportions of the Welch hammer were generous indeed: overall, the hammer weighed about 5,000 pounds, with the anvil alone weighing 1,500. This was comparable to the 60-pound Bradley Rubber Cushioned Helve which weighed about 5,250 pounds, but the Bradley Upright weighed much less: about 3,200 pounds for a 75-pound model.

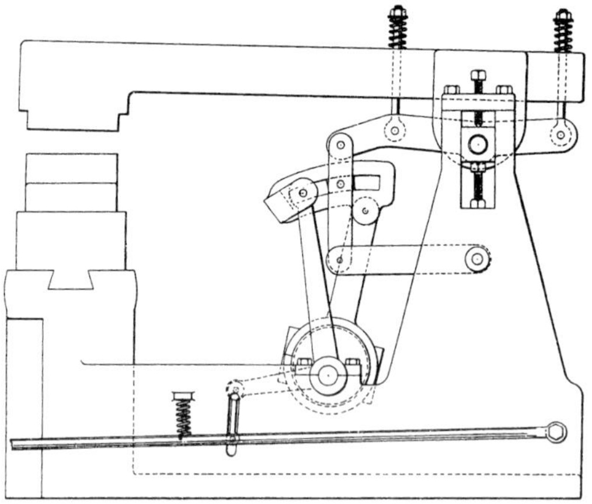

William Welch's patent of 1910 describes an adjustable pitman linkage. This mechanism was capable of bringing the hammer from a stand-still smoothly up to a full stroke with foot pressure control.

The Welch hammer was designed by William H. Welch, who was granted patent number 955,623 on April 19, 1910. The salient feature of the design was an adjustable pitman by which the smith directly controlled the operation of the hammer. The coil-spring-cushioned carriage which supported the helve was connected (via a pivoting link) to a slotted-arm connected to the crank. The pivoting link was attached to a key which moved in the slot. At the extreme forward position in the slot, no motion was communicated to the helve. Motion increased steadily as the link was moved (through foot-treadle control) toward the rear of the slot. This clever approach eliminated the need for a friction clutch or idler pulley. The hammer crank ran continuously, and the hammer came to life whenever the foot-treadle moved the position of the link. Welch's approach was similar to that of Edward Paye more than half a century earlier; however, with Paye's hammer the movement of the linkage could not be set to zero. Another similar design was employed by William Blacker in his Mechanical Striker, which was soon to achieve great popularity and commercial success. The Welch helve hammer, on the other hand, quickly vanished. The Welch Hammer and Machine Company was never listed in the

Welch's Variable Blow Power Hammer. Note the unusual placement of the coil-spring cushions. (*Machinery*)

Thomas' Register of American Manufacturers, and the hammer was never advertised in *The Blacksmith and Wheelwright*. Details of the possible influence of William Welch's design upon the development of the Blacker Hammer, in England, are not known.

* * *

There was one additional coil-spring-cushioned helve to appear on the market during this time. A small, rather simple crank-actuated helve was placed on the market by Fritz A. Schultz, of Chicago, Illinois. The hammer weighed 500 pounds overall, but offered considerable versatility to the metalworker (not necessarily a blacksmith). The anvil could be reversed to present either a flat lower face or a small horn. The head could be fitted with an upper die of either wood, rawhide or steel. Though clearly not a dedicated blacksmith's power hammer, this machine could be used for light forging. It was controlled by a foot-treadle which operated an idler pulley. Little is known about the maker, Fritz A. Schultz; however, a rivet spinning lathe by the same firm is profiled in the *American*

Schultz's versatile metalsmiths' hammer. (*American Machinist*)

Machinist.[110] The hammer never appeared in *The Blacksmith and Wheelwright*; and the company was not listed as a hammer-builder in the *Thomas' Register of American Manufacturers*. The duration of its manufacture is not known.

* * *

The machines built by the Northwestern Power Hammer Company of Primghar, Iowa, brought together two parallel lines of power hammer development. These hammers were belt-driven, crank-actuated helves, but they also incorporated features that were previously found only in foot-powered hammers. Foot-powered hammers, also known by the colorful and enigmatic name "Olivers," pre-date crank-actuated power hammers in American blacksmiths' shops. Their development and use continues to this day.[111] Olivers had certain advantages over belt-driven hammers: they were cheaper to buy or build, cost nothing but sweat to operate, were compact, and often used the smith's existing anvil. Of course, they came up short when heavy or repeated hammering was required; but they were generally designed with careful thought to replacing the human striker in all of his subtle versatility. Toward that end, many Olivers were designed with the capability of adjusting the mechanism to strike at any desired point on the anvil. The idea of combining the capabilities of an Oliver with those of a conventional belt-driven hammer was inspired; and, not surprisingly, it came from a practical blacksmith.[112]

The Northwestern Power Hammer. Smiths were urged to "place this hammer where the old anvil stood." (*The Blacksmith & Wheelwright*)

John F. Overcash applied for a patent on his hammer's mechanism on July 31, 1913. Originally, the design was intended to include "the ordinary blacksmith's anvil."[113] It featured a dual-function control lever with which the smith could move the hammer laterally, as well as back and forth, across the face of the anvil. The hammer's stroke could be precisely adjusted (with the machine in operation) by means of a small crank which moved the connection point of the pitman. Vibration and the shock of impact was absorbed by a beam which extended above the helve, and was connected just behind the hammer head through coil springs. Interestingly, there was no provision to raise or lower the hammer. Therefore, the design suffered from the classic flaw of early helve

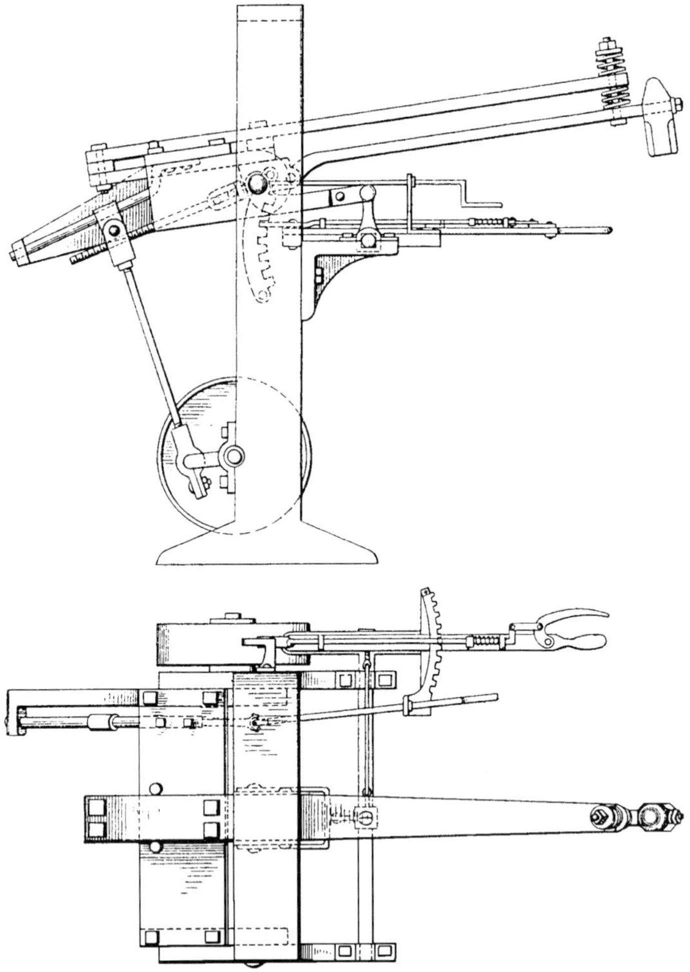

John Overcash's patent of 1914 described the mechanism of the Northwestern Power Hammer: the hand-lever positioned the hammer above and along the anvil. The small crank controlled the stroke.

hammers: the head would strike parallel to the anvil for only one thickness. Patent number 1,103,918 was granted to Overcash on July 14, 1914.

When the Northwestern Power Hammer appeared in the pages of *The Blacksmith and Wheelwright* in the Spring of 1915, some important modifications had been made. The machine included its own anvil, presumably to simplify issues of height, stability and fit. While not exactly traditional in form, the anvil was tailored to suit most of the varied requirements of the smith. The rod to which the pitman was connected now pointed forward, toward the operator, and was adjusted by means of a lever in lieu of the small crank. Most importantly, the hammer head (which was composed of a 12 inch long rod of two and a half inch diameter steel) could be moved up and down at the end of the helve to accommodate thicknesses up to four inches.[114] This unusual arrangement dealt effectively with the issue of keeping the hammer head parallel to the anvil at various heights, though it would have required frequent adjustment.

Advertisements for the Northwestern Power Hammer stressed its versatility. It was pointed out that the head of the Northwestern Power Hammer would reach the anvil, unlike "triphammers" (meaning hammers with a spring linkage, such as the contemporary favorites: the Little Giant, the Star and the Boss, etc.) which would "spring onto the work," requiring two hands to hold the workpiece. This was important, since a blacksmith using one of Overcash's hammers would need a hand free to operate the control lever. In welding a wide wagon tire for instance, it was proposed that the hammer head would be moved back and forth over the lap until it was solidly welded.[115] While this was probably the most convenient manner in which a human striker would be employed to execute such a weld, the logic does not necessarily extend to a mechanical striker. For a power hammer such as a Little Giant, with blows which were dependable in number and location, it would surely have been easier to move the tire itself. In any event, the Northwestern Power Hammer was proclaimed by its builders to be "the only hammer put on the market that is a complete success in welding." It was offered as superior for plow work, pointing and sharpening cultivator shovels, and welding automobile springs. Weighing 1,000 pounds overall, the hammer could be run with a two and a half horsepower gasoline engine.

In January of 1916, Overcash and Son offered the scaled-down Northwestern Power Hammer Jr. True to the original concept, this hammer employed a standard blacksmith's anvil

The Northwestern Power Hammer Jr., with which smiths could "get out of the old trip hammer rut." (*The Blacksmith & Wheelwright*)

(which was presumably not included with the purchase of the machine). The pitman was connected directly to the helve, which wielded a simple, eight pound head. The location of the blow and the stroke were still adjustable, but the head was no longer variable for different thicknesses of work. The Northwestern Power Hammer Jr. was unique in claiming to replace not only the blacksmith's helper and a conventional "triphammer," but the blacksmith's hand hammer as well![116] While it was undoubtedly less expensive than the earlier hammer to produce and sell, it was short-lived; and the company that built it soon vanished.[117]

* * *

John Overcash was not the only inventor to develop a belt-driven mechanical striker. The hammer devised by the Englishman William Blacker is perhaps the best known; inspiring, in turn, the Mechanical Sledge Hammer, built briefly by Chambersburg Engineering Company of Chambersburg, Pennsylvania in the late 1920's. Blacker's design actually pre-dates the Northwestern Power Hammer slightly, though it is not known if Overcash was aware of

The Blacker Hammer, ca. 1928. From this view-point, one can clearly see the slotted-lever linkage. (*The Blacksmith & Wheelwright*)

developments overseas. William Blacker, of Stalybridge, England, applied for a United States' patent on his design September 16, 1912. Patent number 1,054,332 was granted on February 25, 1913; but the hammer was not introduced into the United States until the Spring of 1918. In May of that year, with the First World War at its peak, the Mechanical Hammer Company, Ltd. (Staly Iron Works) took-out a full page advertisement in *The Blacksmith and Wheelwright*, unveiling its hammer. Testimonials on that page reveal that the hammer had been in production for at least five years in England (notably, though, all twelve of those testimonials were from English users).[118] While the Blacker hammer was a British development, and was built in that country for many years, it was apparently also constructed in the United States during the Second World War. For that reason, as well as its influence upon the Chambersburg hammer, it will be described here in greater detail.

Like the Northern Power Hammer, the Blacker Hammer offered the smith a mechanical replacement for the human striker. Blacker's machine appeared in England at a particularly auspicious moment, for brawny blacksmiths and their assistants were soon in demand for front-line combat positions. Such a labor-saving machine was quite simply a god-send, and became an immediate commercial success. More than 1,000 hammers were built in the three years prior to the company's May, 1918, advertisement in *The Blacksmith and Wheelwright.*[119] Though conceptually similar to Overcash's hammer (in providing a mechanism to traverse the hammer across the anvil), the Blacker hammer was very different. The helve was composed of two parallel, pivoting bars, which kept the face of the hammer head parallel to the anvil at all times. The hammer was positioned laterally along the anvil by means of a foot-controlled motor drive. The control pedal could be depressed to the right or left to determine direction. This degree of sophistication in control was typical of British mechanical design, and is reminiscent of Longworth's mechanism for adjusting the fulcrum of his hammer by powered foot control (discussed later, in connection with Arthur Alfred Koch's American version of a movable-fulcrum hammer). There was no provision in Blacker's design to adjust the hammer forward and backward on the anvil. However, the heavy, modified anvil provided with the Blacker hammer was fairly wide, and was presumably deemed adequate to support the workpiece at any point in the hammer's arc. The drive mechanism of the Blacker hammer was also different. Instead of an

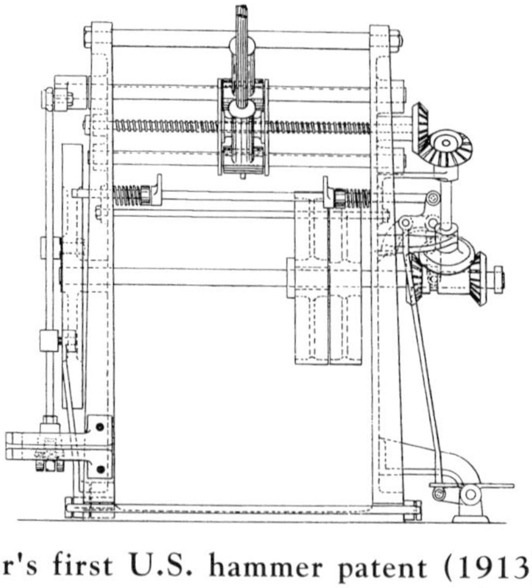

William Blacker's first U.S. hammer patent (1913). Note the reversible bevel-gear arrangement for moving the hammer laterally back and forth under power.

adjustable pitman, which was moved by a hand crank, the foot treadle of Blacker's hammer moved a block along a slotted lever. The crank turned continuously (whether the hammer was actually pounding or not) and the blows started and intensified as the block was moved farther from the lever's fulcrum. This was an extremely clever mechanism, one which had appeared several years earlier in the Welch Variable Blow Power Hammer.

The most arresting feature of the Blacker hammer is the compound, parallel-bar helve which keeps the head parallel to the surface of the anvil at all times. Actually, this design was also not new. In 1862, William D. Grimshaw of Newark, New Jersey, obtained a patent for a hammer which included just such an arrangement. Patent number 35,309 (granted May 20, 1862) was concerned primarily with a novel ratchet and pawl arrangement which allowed the hammer to descend faster than its constantly-moving drive mechanism. The patent claims include the "parallel motion bar," but it is quite possible that this design was already in circulation. It is unknown if Grimshaw's hammer was ever manufactured. In any event, it is surprising that this clever but straightforward design was not more widely used. In 1898,

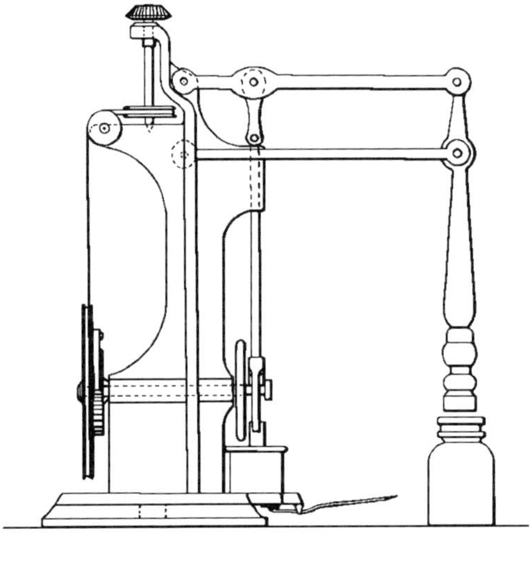

Grimshaw's parallel-bar helve was patented in the United States in 1862, half a century before Blacker's hammer was developed.

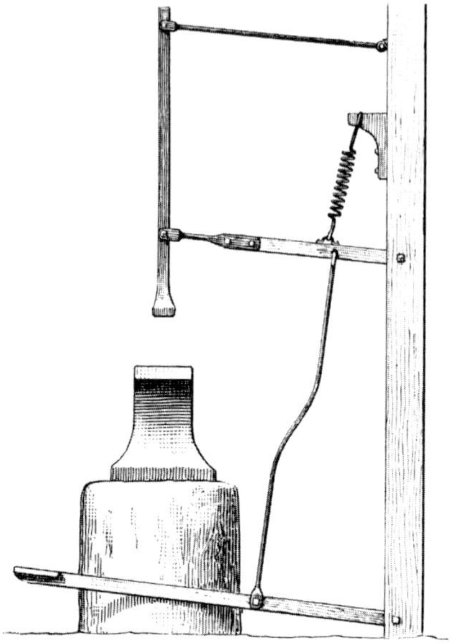

C. J. Palmer's home-made Oliver from 1898. (*The Blacksmith & Wheelwright*)

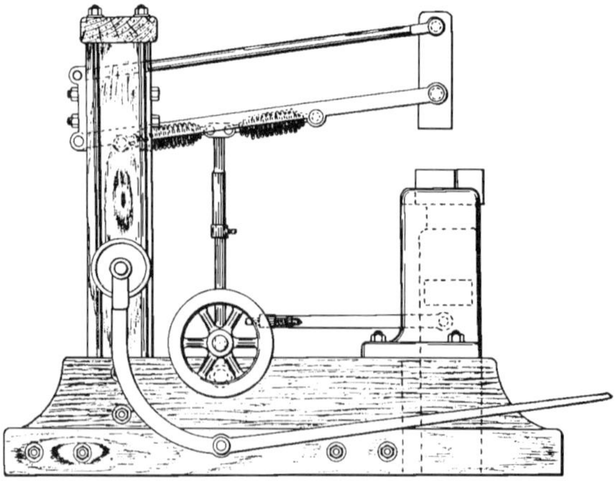

A. C. Gough included an illustration of a parallel-bar helve in an article in 1914. (*The Blacksmith & Wheelwright*)

The Blacksmith and Wheelwright published a drawing of a home-made parallel bar Oliver. Submitted by C. J. Palmer of Wagram, Ohio, the design was as simple as could be imagined, but appeared to be quite functional.[120] A belt-driven version of a parallel-bar helve hammer was illustrated in A. C. Gough's article, "The Power Hammer - It relieves the Smith and His Helper and is a Time Saver," which appeared in *The Blacksmith and Wheelwright* in the Spring of 1914. By that date, though, it is probable that Gough was aware of William Blacker's design.[121]

The idea of a parallel-bar helve, and, for that matter, a hammer which could traverse an anvil, was definitely "in the air" in the early years of the twentieth century. Jean Kunz of Cronberg-in-the-Taunus, Germany, obtained U.S. Patent 1,046,809 on December 10, 1912 for his own variation on this theme. The Kunz design stands in remarkable contrast to the Blacker hammer. Instead of moving the hammer across a stationary anvil, Kunz provided racks and levers which moved the anvil itself both back and forth and side to side. The parallel-bar helve was employed to provide an additional capability: the upper pivots of the parallelogram could be moved to tilt the face of the hammer with respect to the face of the anvil! A well-controlled hammer striking at an angle to the anvil can be extremely useful to the blacksmith,

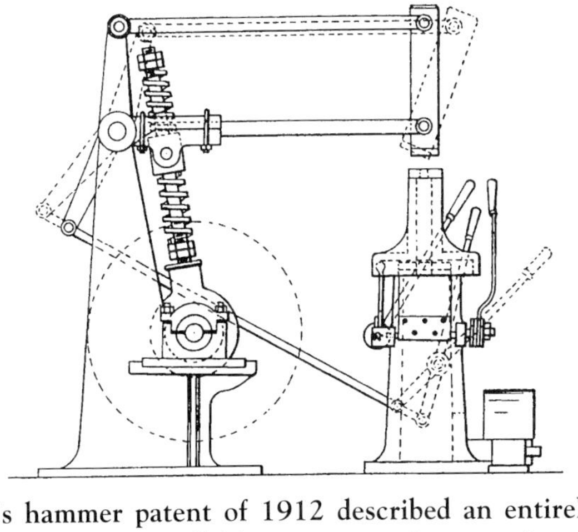

Jean Kunz's hammer patent of 1912 described an entirely different parallel-bar helve: the parallel bars were used to permit precise tilting of the hammer face; also, the anvil was moved about with hand levers.

but previously patented versions of this idea presented a hammer which struck obliquely when tilted, and upon a different spot on the anvil. An adjustable parallelogram resolved these limitations and allowed the hammer to be adjusted while in use. The extent to which Kunz's hammer was developed is not known.

The anvil provided with the early Blacker Hammer weighed 500 pounds, and was conventional with the exception of a cut-away portion in the front. This detail was provided so that the forward edge of the anvil could be aligned with the edge of the hammer head, permitting a perfectly square double shoulder to be drawn-down. The machine weighed 1,100 pounds overall, and was extremely efficient, requiring only one horsepower to run. With a working space of eight and a half inches in height, practically any existing hand tools (upper and lower swages, etc) could be used. The hammer was said to be ideal for stock up to two and a half inches square.[122]

Advertising for the Blacker Hammer was both imaginative and effective. (*Blacker Engineering Co. trade literature*)

Advertisements in *The Blacksmith and Wheelwright* from the Spring of 1918 solicit "propositions of every kind." Unfortunately, this was a difficult time to conduct business across the Atlantic In September, a notice appeared asking readers to repeat their responses to the Mechanical Hammer Company, since several ships bearing the mail had fallen prey to German submarines![123] It was

POWER BLACKER HAMMERS SLEDGE

not until after the war that the Blacker hammer became established in the U.S. market. The Blacker Engineering Company was incorporated in the United States (located in Grand Central Terminal in New York City), and the hammer was profiled in the pages of *American Machinist* in January of 1923. Redesigned slightly, the hammer now weighed 1,300 pounds overall, with a 560-pound anvil. The special base for the anvil weighed another 336 pounds. The hammer could be purchased with 15 inch diameter tight and loose pulleys for belt drive, or it could be had with its own electric motor, geared to the drive shaft.[124] A new profile of the Blacker hammer, in *The Blacksmith and Wheelwright* from the Summer of 1923, noted that it was the only hammer then available that could be geared direct for electric drive. It also noted that the first U.S. sales of the hammer dated from mid-summer of the previous year, but that the machine had already been adopted "as standard equipment by many of our largest and most discriminating concerns... Pennsylvania Railway; Long Island Railway; Baldwin Locomotive Works; Electric Manufacturing Company, Pittsburgh; the National Tube Works of the U. S. Steel Corporation and many others."[125]

William Blacker had previously obtained two additional United States' patents: number 1,297,990 (granted March 25, 1919) and number 1,349,939 (granted August 17, 1920). The former patent describes a mechanism for raising and lowering the hammer, and the latter describes an improved means of providing a smoothly-reversing lateral movement of the hammer across the anvil. The Blacker hammer was promoted strongly in the United States in 1925, shortly before the expiration of Blacker's patents. Faced with competition from Chambersburg's Mechanical Sledge, which appeared in 1926, the Blacker Model "B" was placed on the market in 1928. It was claimed that more than 4,000 Blacker hammers were then in operation around the world.[126] It is unknown if any of these Blacker hammers were built outside of

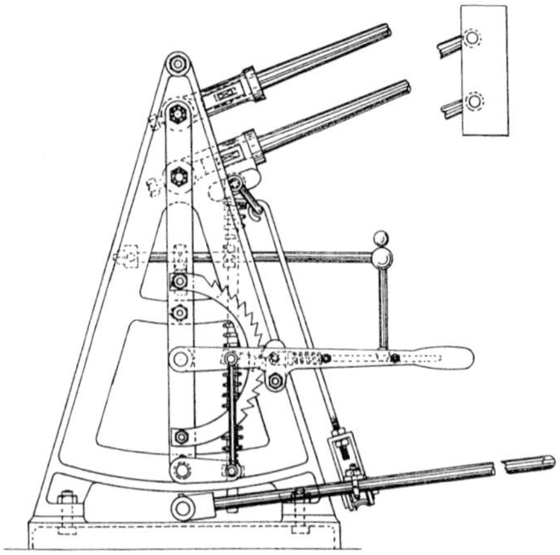

William Blacker's second U.S. patent (1919) described a mechanism for raising and lowering the parallel-bar helve.

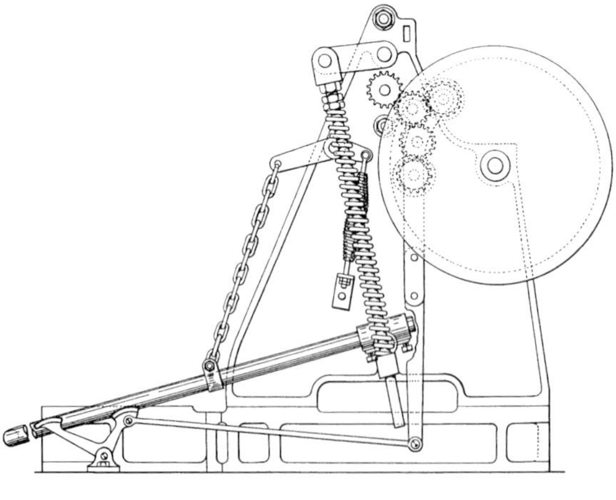

William Blacker's third U.S. patent (1920) was concerned with a smoother reversible power traverse.

Britain. However, by 1944 there had evidently been some manufacture of these hammers in the United States. On July 5, 1944, the McKiernan-Terry Corporation of New York City, announced that it had purchased all of the physical assets of the Blacker Engineering Corporation, including inventory, records, drawings, patterns, jigs, tools, and fixtures. The announcement continued: " The Blacker Hammer henceforth will be manufactured in our Dover, N. J. plant where for forty years the world's leading Double Acting McKiernan-Terry Pile Hammers have been built..."[127] The ultimate fate of the Blacker hammer in the United States is not known. It was manufactured by McKiernan-Terry at least through 1945, but it was listed in the *Thomas' Register of American Manufacturers* for only that one year. However, the hammer continued to be built in England (perhaps as late as the 1980's), and a virtually new Blacker Model "B" hammer was displayed at the conference of the Artist Blacksmiths' Association of North America held in San Luis Obispo, California, in 1992.[128]

* * *

The success of the Blacker hammer in the United States and abroad stimulated competition from an unlikely quarter. The Chambersburg Engineering Company of Chambersburg, Pennsylvania, was (and is) one of America's most important builders of closed-die forging machinery. In addition to board and steam drop hammers, the company built single and double-frame open-die hammers, as well as self-contained pneumatic hammers. The history of this manufacturer and its machines will form an important chapter in a volume dedicated to steam- and air-actuated power forging hammers, but it deserves mention here. From 1926 through 1929, the Chambersburg Engineering Company built a Mechanical Sledge which resembled the Blacker Hammer. Though its manufacture was quickly abandoned, the Chambersburg Mechanical Sledge was extremely robust, and examples have survived to this day.

The Chambersburg Engineering Company was founded in 1897 by William H. Derbyshire. He had left the prominent Bement-Miles Company of Philadelphia to manufacture his own heavy forging machinery. Having acquired the plant of the Taylor Manufacturing Company in Chambersburg, Derbyshire's first hammers resembled the single and double-frame steam hammers that he helped to build in Philadelphia. Board drop hammers were introduced in 1912, and became the company's principal product.

The Chambersburg Mechanical Sledge was a more robust, but less sophisticated, version of William Blacker's hammer. (*Chambersburg Engineering Company - The First Fifty Years*)

The 1920's were a time of intense research and development. Production and profits slowed with the post-war economy, and the Mechanical Sledge marked an attempt to broaden the company's base in the forging tool market.

The Mechanical Sledge first appeared in the pages of *Machinery* in the summer of 1926. Shown from its right-hand side, the machine greatly resembled the Blacker Hammer. Like the Blacker, it offered a head which was always parallel to the anvil, but it claimed that the power traverse was activated by the same treadle which was used to control the blows. The hammer head weighed 50 pounds, and could strike a maximum of 218 blows per minute (a bit faster than the Blacker's 140). The hammer required two and one half horsepower to operate (significantly more than the Blacker).[129] In early 1928, a revised version of the Chambersburg Mechanical Sledge was unveiled. The powered traverse of the hammer across the anvil was now controlled by a hand lever, and the construction was said to be heavier. Complete with its motor, the machine weighed a substantial 4,500 pounds![130] An illustration of the left-hand side of the Mechanical Sledge, in *Chambersburg Engineering Company - The First Fifty Years*, reveals a significant difference between the Chambersburg Mechanical Sledge and the Blacker hammer. Since William Blacker's slotted-lever drive design was still protected by patent, the engineering department at Chambersburg employed the old mechanical standby of a slipping belt and idler pulley.[131] There were no patents associated with the Chambersburg Mechanical Sledge. An internal memo detailing improvements upon the Blacker design and Chambersburg's early efforts was signed by Howard Terhune, Chambersburg's chief designer, but pursuit of this project was abruptly curtailed.[132] The Mechanical Sledge never represented more than 2% of the company's receipts for any of the years in which it was in production. Development of Chambersburg's revolutionary "rigid frame" drop hammers justly became the new focus of its attention.[133]

While the Northwestern Power Hammer and the Chambersburg Mechanical Sledge Hammer were relatively short-lived, the Blacker hammer has endured into the last decade of the twentieth century. So too have a number of various foot-powered hammers. The blacksmith's profession has changed dramatically since the early years of the century, but several constants remain: the smith still requires occasional help at the anvil, and the cost of human labor remains high. The direct-acting air-powered Kick-ass Striking Hammer, designed by Grant Sarver and built by the Apex Forge and Tool Company of Tacoma, Washington, represents a new direction in this type of hammer. Just as many early power hammers appeared with the new availability of gasoline engines, Sarver's hammer takes advantage of the small compressors which have become ubiquitous in the modern blacksmith's shop.[134] The challenge of economically and effectively replacing the human striker will undoubtedly continue to spur new power hammer invention well into the twenty-first century.

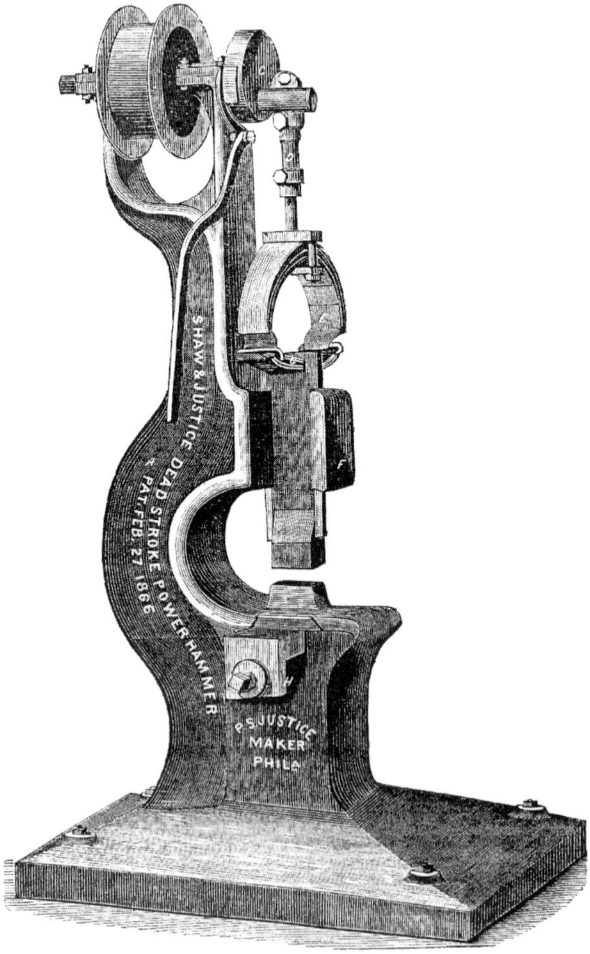

The Shaw & Justice Dead-Stroke Power Hammer used a bow-shaped, metal spring linkage. The ram was suspended from a leather strap, further isolating the hammer's mechanism from the impact. (*The Scientific American*)

CHAPTER 4

"The Helper with No Bad Habits"

Perhaps the most successful crank-actuated open-die power forging hammer type was vertically-configured with a guided-ram. Tens of thousands of these machines were built in America following their first appearance immediately after the Civil War. In sizes up to 500 pounds (and even, in rare instances, 1,000 pounds), they were effective forging machines and were particularly efficient in their use of valuable shop space. A surprisingly large number of these hammers - more than any other crank-actuated type - are still in service. When worn, they are painstakingly restored; and they continue to earn their keep, both as labor-saving devices and artists' tools. A blacksmith today might purchase a new self-contained air hammer (foreign-made), but a second-hand vertically-configured crank-actuated hammer is often more economically-accessible. While no new machines of this type are currently being manufactured, it is still possible to purchase replacement parts for the ones most widely used.

Vertically-configured crank-actuated hammers appeared within a decade of the American development of atmospheric hammers. In fact, Milo Peck had suggested that his atmospheric hammer could be arranged with the crank placed directly above the air-cylinder and ram. This was hardly a radical idea, for drop hammers with Peck's patent crank lifter had been so-configured since the early 1850's. In any event, early atmospheric hammers such as those designed by Bennet Hotchkiss were not vertically-configured. This was most likely due to a desire to maintain a low center of mass for maximum stability. Locating the crank and drive shaft below the atmospheric linkage also made it possible for the machine to fit into the most confined spaces. The air-cylinder linkage was not as compact as Shaw's bow-shaped steel spring, and overall height was probably a consideration. Nevertheless, it was possible to build a vertically-configured crank-actuated hammer with an air-cylinder linkage. Such a design was developed in England and manufactured in the United States, as the Hackney hammer, during the early 1890's. The re-introduction from abroad of a fundamentally American development in forging machinery was remarkable, but it was also apparently short-lived.

A vertical configuration for a crank-actuated hammer made sense both mechanically and commercially. The compact, straightforward arrangement required a minimum of parts, thereby reducing the costs of manufacture and the final sales price. Vertically-configured hammers shared an important design advantage with the larger (and far more costly) direct-acting steam hammers for they also used guided-rams which maintained constantly-parallel die faces. This arrangement made these machines far more versatile than the helve hammer. It simplified the working of various thicknesses and the interchanging of dies. The simplicity of the vertical configuration made it popular with those blacksmiths who designed and built their own hammers; and the majority of "home-made" hammers, illustrated over the years in the pages of *The Blacksmith and Wheelwright*, were of this type.

Vertically-configured hammers were effective and efficient forging machines. Inspired initially by Hotchkiss' air-cylinder spring linkage, an endless variety of spring linkages were developed, using metal springs or rubber cushions. These linkages served to isolate the hammer and its drive from the shock of repeated impact, and were also capable of generating an impressive blow in a very compact space. The upward momentum of the ram was stored by the spring and used to deliver a blow of much greater force than would have been possible with gravity alone. Many of the linkages that were developed for vertically-configured hammers subsequently appeared in crank-actuated helves. In fact, the development of crank-actuated guided-ram hammers likely spurred the further development of the venerable helve. Guided-ram hammers held certain conceptual advantages over trip-actuated helves, but the comparison was not that simple. The first vertically-configured crank-actuated hammer, designed by Thomas Shaw, was offered as the "Dead-Stroke" Power Hammer. This term referred to the fact that all of the energy of the descending hammer ram was transferred to the work before the crank and its linkage lifted it upward. While a great deal of forging can be accomplished with such blows, a lighter touch is often desired. The hammer blows from a rapid trip-actuated helve, or a crank-actuated helve such as the Bradley Rubber Cushioned Helve Hammer, were not necessarily "dead." With these machines it was possible to obtain a springy blow in which the hammer ram was lifted quickly off of the workpiece before a dead blow could be delivered. This sort of blow was particularly effective in working delicate, thin parts, as in cutlery manufacture. Such blows were necessarily delivered rapidly, with less cooling of the work between blows. Light, rapid blows were preferred for many swaging operations, and they allowed fine control of the structure of the metal. The subtle qualitative peculiarities of the hammer blows delivered by these different mechanical designs helps to explain their diversity, and the persistence of seemingly antiquated types.

During the latter half of the nineteenth century, a number of different companies built vertically-configured, crank-actuated hammers. While some continued to build a full-range of hammers well-into the twentieth century, the market for these machines was changing. The early decades of the twentieth century witnessed a boom in power hammer use, as smaller forges obtained power. During these prosperous decades, and particularly during the years of the first World War, smiths were offered a dizzying array of hammers, including many vertically-configured models. The most enduring of these were the hammers built by the Mayer Brothers, who started out in Mankato, Minnesota, in the 1890's. The Mayer Brothers' Little Giant was originally produced for the new, turn-of-the-century hammer market, but eventually grew to dominate the manufacturers' hammer market as well. So completely did the Little Giant command the steadily dwindling hammer market of the mid-twentieth century, that many people forgot about the earliest vertically-configured, crank-actuated power forging hammers and the man who first employed a metal spring linkage.

Thomas Shaw was a native of Philadelphia, born in 1838. He became a machinist's apprentice at age 16, and rose to become one of the most respected American engineers of the latter nineteenth century. Granted more than 186 patents, covering a wide range of invention, he is perhaps best remembered for his development of instruments to detect and signal the presence of toxic gasses in underground mines. He also invented the spring lockwasher, which was of immediate, worldwide importance to railroad engineers.[1] Before establishing his own engineering firm in 1872, Shaw worked for the respected Philip S. Justice Company, which specialized in "railroad supplies." During his tenure with this firm, Shaw applied himself to, among other things, the use of semi-elliptical springs for railroad cars. In drawing-out the ends of these flat spring plates, Shaw became increasingly aware of the short-comings of the trip-actuated helve hammers. In a most logical step, he devised a spring linkage for a crank-actuated power forging hammer which used a semi-elliptical (or "bow-shaped") multi-leaf steel spring. This linkage, with its leather straps, effectively isolated the hammer from the impact and delivered a powerful "dead-stroke."

The first working model of Shaw's hammer was a 60-pounder, and it was put to work forging spring leaf ends in the Justice shop.[2] Shaw was granted U.S. patent number 52,894 on February 27, 1866, and by the end of the year his hammer was being produced in eight sizes, from 15 to 1,000 pounds.[3] The heavier sizes were built in a non-vertical configuration, employing a walking beam which was oscillated by a crankshaft via a long connecting rod. An article on the front page of *The Scientific American* on December 8, 1866, proclaimed the company's intention to build them in sizes larger still, up to ten tons![4] While it is highly unlikely that such immense machines were ever built, Shaw was granted a second patent, number 84,221 on January 17, 1868, intended to protect the walking-beam design (as well as a brake and a more sophisticated means of connecting the ends of the elliptical spring to the leather straps which held the ram).

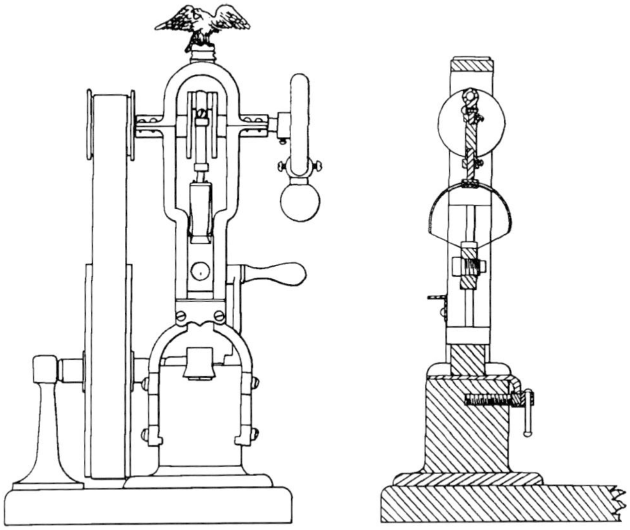

Thomas Shaw's patent of 1866 was illustrated with a drawing of a hand-cranked model. The actual hammer used a crank-plate rather than a crank-shaft. Note the plug for adjusting the weight of the ram.

The *Dictionary of American Biography* mistakenly credits Thomas Shaw with the invention of a steam hammer.[5] He did, however, make another fascinating contribution to the field of open-die power forging hammers. Shortly after developing the Dead-Stroke Power Hammer, Shaw designed a gun-powder pile-driving hammer. A demonstration of this device on June 14, 1869, at the Philadelphia shipyard of John Lynn, was reported in the British press as an unqualified success. A pile was driven ten feet six inches through "stiff, hard soil," in fifty-five blows taking only one and a quarter minutes.[6] This controlled use of what was generally considered an unruly substance, brought another application to Shaw's mind. In England, James Nasmyth's steam hammer had inspired a pile driver. Shaw's gun-powder pile driver, conversely, spawned a forging hammer! The hammer was set in motion by its operator, who tossed a packet of white gunpowder into a cylinder and released a ram to fall from above. The ram entered the gun-powder cylinder and exploded the powder, propelling the cylinder (and the hammer-ram below it) downward

Thomas Shaw's gunpowder pile-driver in action. (*The Masterpieces of the Centennial Exhibition*).

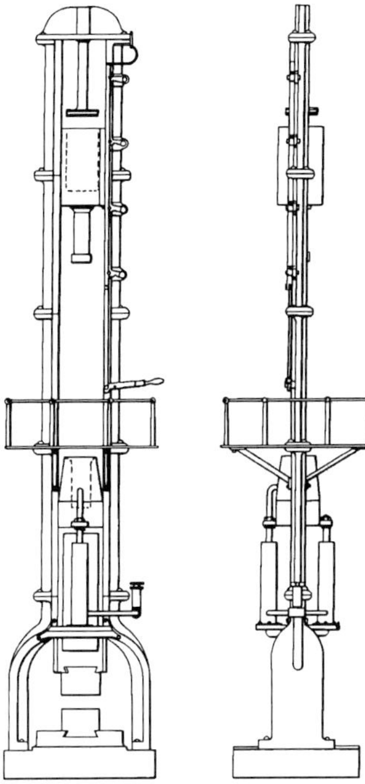

Shaw's gunpowder forging hammer, patented in 1870 (#99,958).

to strike a blow to the workpiece. At the same time, the upper (firing) ram was returned to its perch, ready for the next shot. A curiosity, this hammer left no significant mark on the history of forge hammers.[7] However, the pile-driver was popular for a time, and attracted great interest at the 1876 Centennial Exhibition in Philadelphia.[8]

The Philip S. Justice Company manufactured hammers of Shaw's design in Philadelphia for almost half a century. The firm was listed under "power hammers" in the *Thomas' Register of American Manufacturers* through 1915. Testimonials printed by the Philip S. Justice Company describe the many, varied applications to which the hammer was put and its great efficiency. H.H. & S.F. Hansell, makers of table cutlery in Philadelphia, offered to exhibit the Dead-Stroke Power Hammer at work in their factory alongside a modern tilt hammer. They wrote: "Two men on the tilt hammer averaged about fifteen hundred blades forged per day. One man on your 50-lb. hammer averaged about twenty-five hundred per day." In the same factory, a twenty-five pound Shaw & Justice hammer, used to forge tangs continuously for six months, cost only $5 in maintenance.[9] The Southwark Foundry, also of Philadelphia (which had licensed and manufactured several British types of steam hammers), placed a 50-pound Dead-Stroke Power Hammer in service forging bolts:

> "Our custom is to draw bolts from square rods and then finish the heads to a swage. Hitherto the drawing has been done in an 'Oliver;' a good day's work for one man and his machine being 200 bolts, say 5/8 to 3/4 by 2 1/2 to 3. With the hammer, one man can draw at least 800 bolts a day and two other men can finish the bolts. Thus three men and a hammer make as many bolts as four men and four 'Olivers.'"

The superintendent of the Hudson River Railroad expressed satisfaction with his 1,000-pound hammer, remarking upon its ability to forge work from a half inch to seven inches in diameter without adjustment.[10]

* * *

Well before Thomas Shaw's patents lapsed, manufacture of the Dead-Stroke Power Hammer was licensed to other builders. One of these firms was the Hull and Belden Company of Danbury, Connecticut. In 1876 it was building six sizes of the hammer (all vertically-configured), and had shipped machines to England, Russia, Switzerland, New Granada, Peru and India.[11] The hammer was quickly improved by the addition of an adjustable crank pin. The pin could be moved within a T-slot across the face of the crank plate, allowing for greater or lesser throw. The company subsequently became the R.A. Belden Company, and from 1886 was a regular advertiser in *American Machinist*. By this time, significant changes had been made to the hammer ram and its guides, which were now oriented front to back. This change was intended to stabilize the ram, making the hammer more suitable for work using swaging dies. Russell Belden observed that, despite the virtues of Shaw's design, a certain measure of "jar and shock...has heretofore

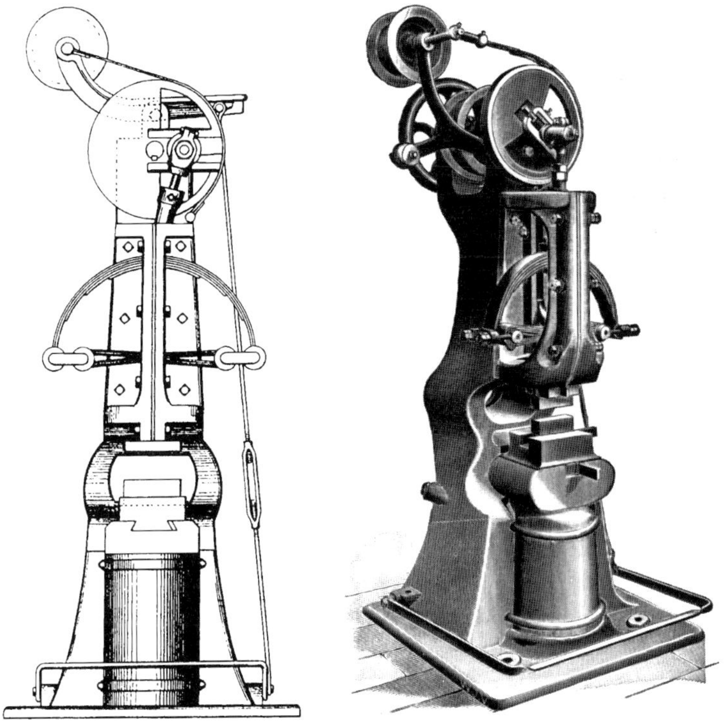

Russell Belden's improvement featured long guides, which stabilized the spring as well as the ram. Belden's machine ultimately became the Scranton Improved Upright Power Hammer. (*The Iron Age*)

been unavoidable in this class of hammers."[12] Indeed, when the Bradley Company of Syracuse, promoted its own helve hammers, it was quick to point out that the play required in the guides of the dead-stroke hammer (to allow for the expansion of the ram as it heated-up in running) "allowed the ram to run loose enough to shuckle...fatal to nice die swaging."[13] This modified hammer was known as the Belden Improved Power Hammer, and the company prospered in Danbury until June 18, 1888, when a dramatic fire destroyed their drop forging shop and machine works.[14] Within several months the company had found a new home, relocating to New Haven, Connecticut. There, it took over the large shops previously occupied by the Mansfield Elastic Frog Company.[15] At that point the firm also changed its name to the Belden Machine

Company. A large drop-forging shop was assembled, and it continued to manufacture the improved hammer (in sizes from 20 to 150 pounds), as well as a crank planer. Russell Belden's improvements to the Shaw hammer were granted U.S. patent number 403,453 on May 14, 1889. These included the previously mentioned orientation of the ram and guides, as well as a means of shimming the anvil to protect the machine from the shock of the hammer's impact. Another unusual feature of Belden's design was the great length of the guides, which guided not only the ram, but a slide that secured the bow spring. This arrangement restricted and focused the movement of the linkage, which helped to stabilize the hammer. Despite these improvements, or perhaps because they permitted a higher speed and harder blows, the company determined that certain parts were not of adequate strength. Early in 1889, the hammer was redesigned and an entirely new set of patterns was made.[16] At some point prior to the end of the century, the manufacture of the Belden hammer was assumed by Scranton & Company, of New Haven. The hammer became known as the Scranton Improved Upright Power Hammer, and was built under that name until the beginning of the Second World War.

* * *

Another company which licensed the technology of the Shaw and Justice Dead-Stroke Power Hammer was Williams, White & Company of Moline, Illinois. It built its first hammer of this type in 1873 and manufactured them for more than half a century.[17] The Williams, White hammer was always referred to as the "Justice Hammer," despite acknowledgment in its early trade literature that Thomas Shaw was the inventor.[18] Like Russell Belden, Williams, White & Co. made improvements to Shaw's design, adding a slot across the crank-plate to allow for adjustment of the throw. However, it did not depart from the general design of the ram and guides first employed by Shaw. After the turn of the century, a novel modification of the spring linkage became available as an option. This so-called "spring-yoke outfit" was patented by

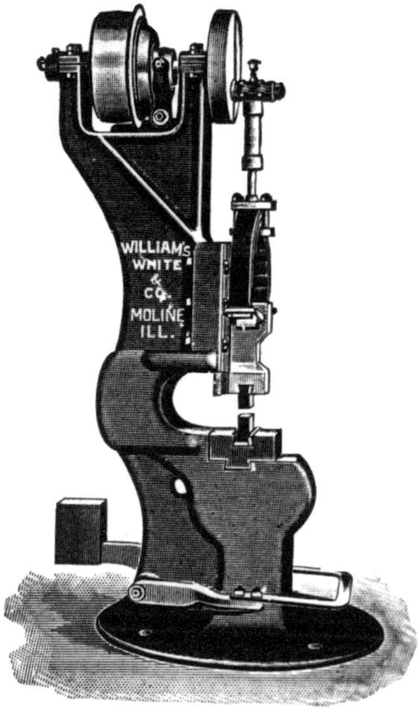

The Williams, White Justice Hammer with a silico-manganese steel spring and double friction clutch. (*Williams, White & Co. catalogue*)

John Norquist in 1904 (patent number 753,618 granted March 1). Instead of a bow-shaped leaf spring, it employed springs of coiled steel or rubber, which were mounted atop a yoke. The toggle arms pivoted on the yoke, compressing the springs, and providing an action very similar to that of the simple leaf spring. This arrangement was apparently favored by some users, but the company continued to supply its hammers with a silico-manganese bow-shaped spring, and recommended this linkage as superior.[19] An arrangement similar to that of the Williams, White spring-yoke outfit can be found on the hammers built by C.C. Bradley & Son, and Kane and Roach, both of Syracuse, New York. The Williams, White Company line of power hammers was further distinguished by a selection of hammers designed specifically for welding steel tires. In lieu of the regular anvil, these hammers employed a lighter,

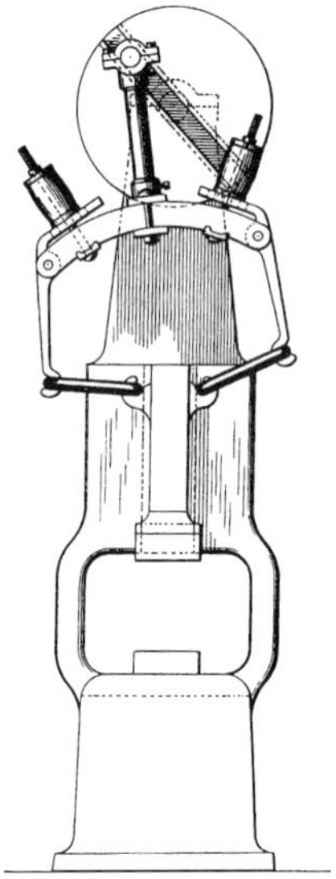

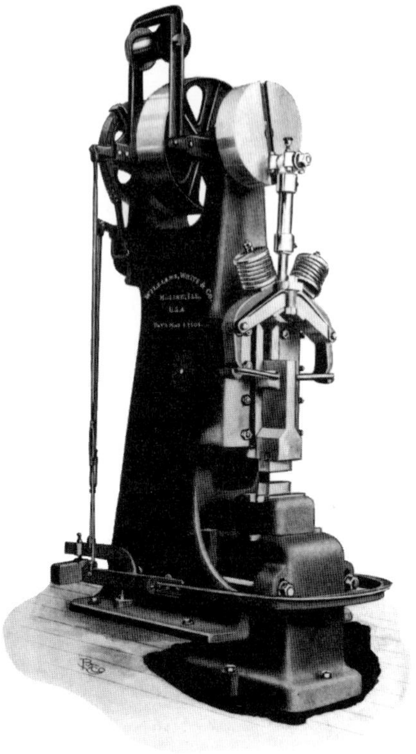

John Noquist's "spring-yoke outfit." Rubber cushions appear in the patent drawing, while a heavy Williams, White Justice hammer shows the design with steel coil springs. Below, a Williams, White tire-welding hammer: note the additional horizontal hammering mechanism. (*Williams, White & Company catalogue*).

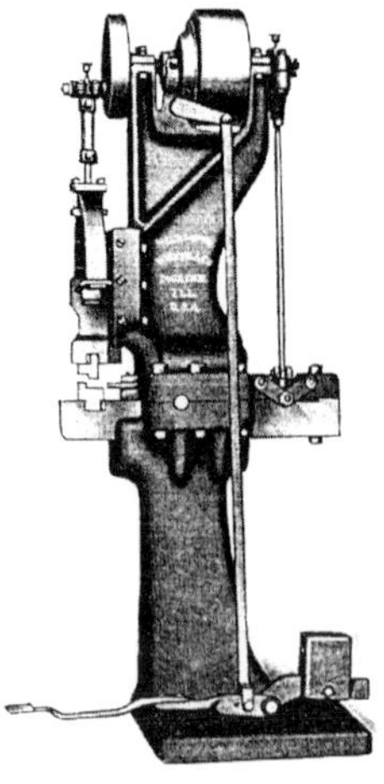

horizontally-oriented anvil bar. Intended for welding modern steel-spoked wheels, these hammers could weld the tire either before or after spoking. Some of these machines were equipped with a curious device (operated by an eccentric on the back end of the drive shaft) for imparting an additional side-stroke.[20]

* * *

While many open-die power forging hammer developments occurred abroad (stimulating inventors in America to seek variations), the Shaw & Justice Dead-Stroke Power Hammer was hailed as a true example of Yankee ingenuity. At least one British inventor found a challenge in Shaw's spring linkage: John Patterson of London, obtained U.S. patent 244,927 dated July 26, 1881, for a linkage that employed a stout, bow-shaped yoke with toggle-arms that pivoted in blocks which held flat coil springs. The reason stated for this design is both interesting and curious. Patterson wished to eliminate the single or multiple leaf bow-shaped springs because of their tendency to fail under repeated, alternating tension and compression. He pointed out that only the best materials worked with the greatest of care could produce a spring that would stand up for any length of time. He further observed that the delays caused by such breakage, as well as the weight and expense of obtaining duplicate springs, "has prevented in some measure the extensive use of the hammers becoming general in mountainous or inaccessible parts of the country."[21]

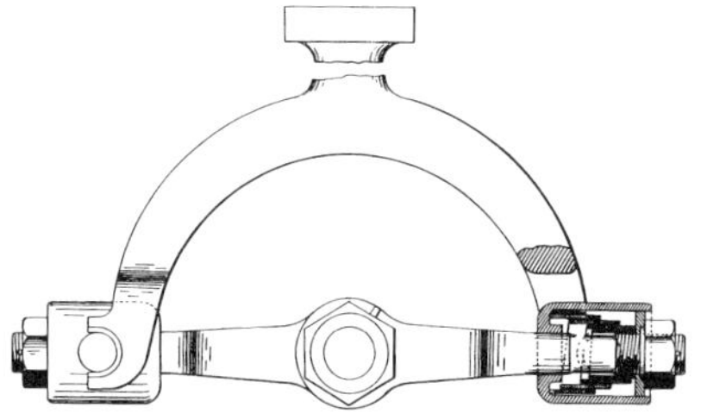

John Patterson's flat coil spring linkage was patented in the United States in 1881. The springs were located in cups which pivoted at the ends of a non-flexing yoke.

It is highly unlikely that Patterson's design was ever incorporated into actual hammers for out-of-the-way and hard-to-reach places, but it is interesting to note that the breakage of springs in early vertically-configured hammers was common enough to inspire his efforts.

* * *

Yet another American company built the Dead-Stroke Power Hammer. The firm of Dienelt and Eisenhardt, of Philadelphia, manufactured its hammers from at least as early as 1881. At that time the company manufactured power hammers in addition to hydraulic jacks and a variety of textile machines.[22] For many years its hammer line was carried by Manning, Maxwell & Moore, of New York, in eight sizes, from 15 to 1,000 pounds, as well as four sizes of so-called "bracket" hammers. The latter machines, were built in 5-, 15-, 25- and 50-pound sizes, and were designed to be mounted to a post or wall, permitting the use of various and/or movable anvils. They had a shorter stroke and were "suitable for cold pounding, planishing, etc., and are a wonderfully handy tool

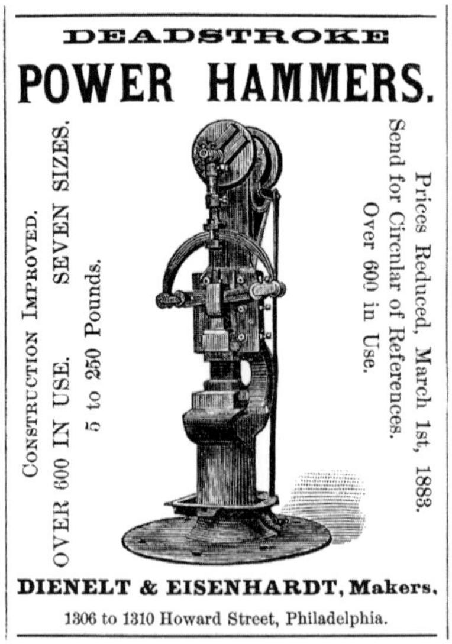

An advertisement for Dienelt and Eisenhardt hammers. (*American Machinist*)

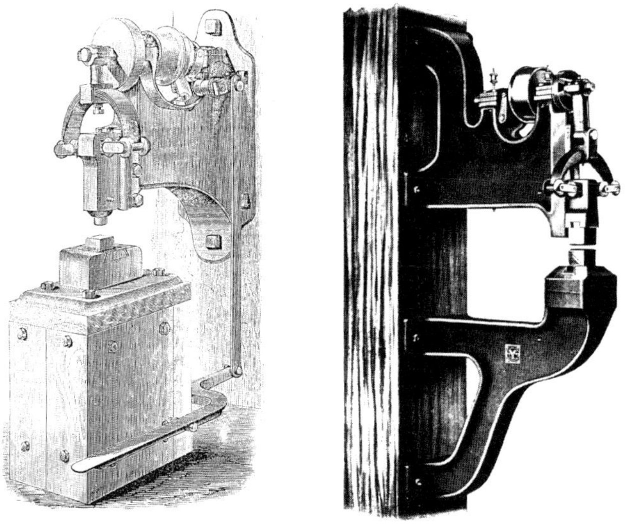

On the left, a Dienelt and Eisenhardt bracket hammer from 1894 (*Manning, Maxwell & Moore catalogue*). On the right, a Yoder K-90 Post Hammer from the 1920's. (*Yoder Company catalogue*)

in any shop."[23] These hammers were not primarily intended for forging, which is also true of similar hammers built later by Yoder, Quickwork, and Pettingell. Hammers of this "bracket" type remain in use today, and are highly prized by specialists in sheet-metal forming.[24] Dienelt and Eisenhardt continued to advertise its hammers through the middle of the twentieth century.

* * *

Thus, in the 1880's and 90's there were at least four different companies building vertically-configured, crank-actuated, guided-ram hammers patterned directly after Shaw's design. Despite financial crises in 1873, 1882 and 1893, the demand for this type of machinery soared. The availability of cheap steel coupled with the demands of rapid growth meant prosperity for America's forges. Hammers of greater efficiency, economy and durability were sought by shops producing such goods as carriage hardware, agricultural implements and cutlery. This ever increasing demand inspired inventors and businessmen to develop and market new variations of this highly successful type of hammer. In 1888, a year before Russell Belden patented his improvements to the Dead-Stroke Power Hammer, Henry St. Lawrence of Northampton, Massachusetts, was granted a patent for a distinctively different design. St. Lawrence's hammer soon become the popular Howe Power Hammer.

Henry St. Lawrence was Canadian-born, of French parentage. He arrived in Northampton in 1883, and assumed the position of foreman in Charles H. Maynard's Hoe Shop. Credited with improving the machinery there, he concurrently obtained patents for his developments.[25] On February 2, 1867, he was granted patent number 335,394 for a non-vertically-configured hammer with a flexible beam (similar to Joseph Palmer's hammer, but having the pivot at the end, rather than the mid-point, of the beam). There is no evidence that this hammer was produced for use

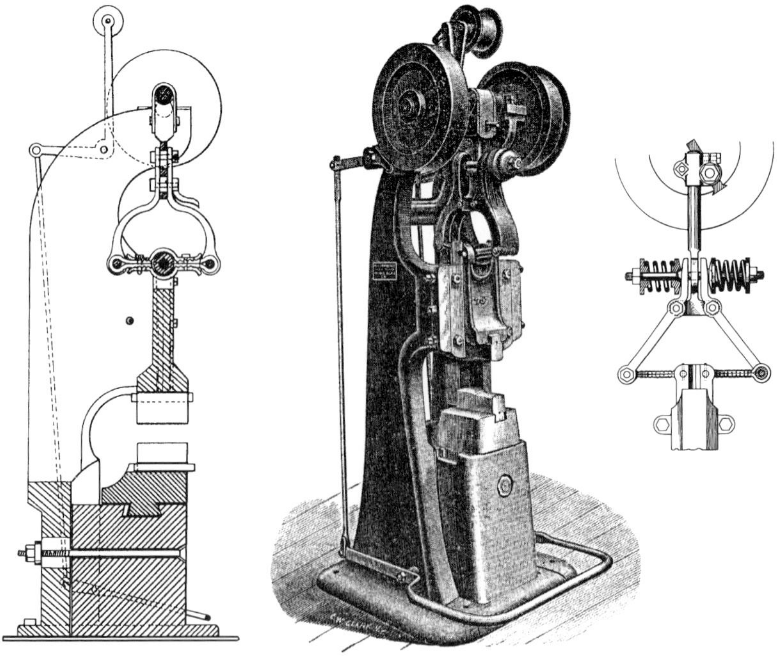

The Howe Power Hammer was based upon Henry St. Lawrence's 1888 patent (left). However, the hammer's distinctive spring linkage was patented by Leonard D. Howard in 1894 (right). (Engraving courtesy of the Bailey/Howe Library, University of Vermont)

beyond Maynard's Hoe Shop. A second hammer patent, number 379,340 was granted on March 13, 1888, and describes the machine which became the Howe Power Hammer. It is likely that the hammer was initially built by (or for) the inventor himself, for a city directory from the late 1880's refers to St. Lawrence as "patentee and manuf. of power hammers and forging machines."

St. Lawrence's design employed a robust frame which supported a crank shaft rather than a crank plate and pin. Shaw's elliptical spring was replaced by a yoke, with improbable "flexible strips" to impart the motion to the ram. Rights to the hammer were soon acquired by the Howe Scale Company of Rutland, Vermont, but the linkage was replaced with one using pivoting arms which compressed rubber springs. A patent describing this linkage (number 513,326) was granted to Leonard D. Howard of St. Johnsbury on January 23, 1894. The Howe Power Hammer was offered in five sizes, from 25 to 150 pounds, priced from $200 to $450.[26] The Howe Scale Company continued to produce and advertise its distinctive hammer through the First World War.

* * *

At about the same time that St. Lawrence and Belden were refining their hammers, Philippe Denery Dupont patented his important variation of the Shaw spring linkage. A Canadian residing in St. Johnsbury, Vermont, Dupont was granted patent number 429,807 on June 10, 1890. He devised an extremely simple linkage that used a single, short coil spring which was compressed by two pivoting toggle arms. This configuration was immediately successful; and was, in fact, the design employed by the Mayer Brothers at the turn of the century in their redesigned Little Giant hammer. The Dupont hammer was first produced in relatively small numbers at the foundry of O.V. Hooker & Son, in St. Johnsbury.[27] Several of these early models were placed in service in the scale works of St. Johnsbury's primary manufacturer, E.T. Fairbanks and Company.[28] The hammer quickly attracted the attention of C.C. Bradley, of Syracuse, who offered to buy the design in February of

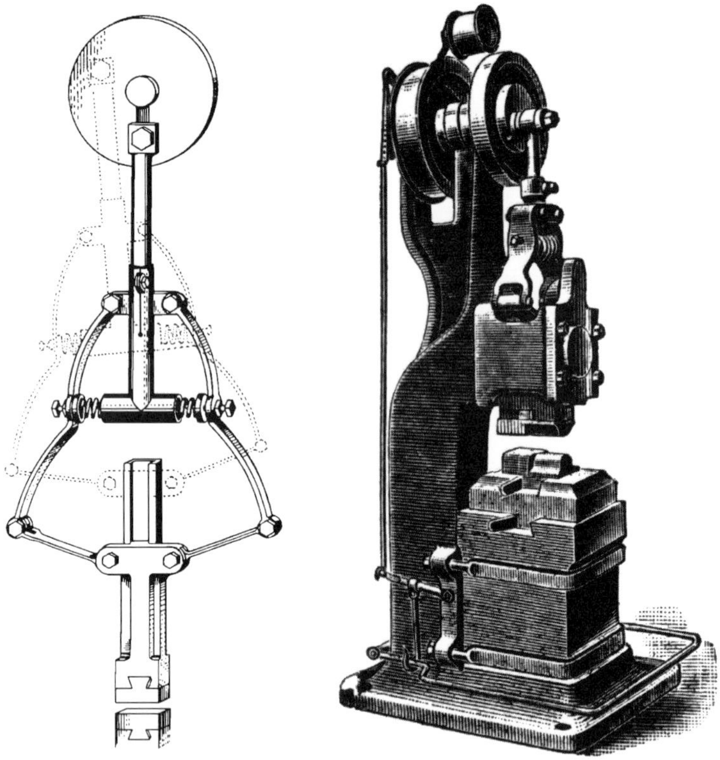

Philippe Dupont's influential linkage was patented in 1890. The Dupont Power Hammer became the Fairbanks Power Hammer in 1902. (*The Blacksmith & Wheelwright*)

1891.[29] Dupont refused to sell and the Dupont Manufacturing Company was formed in early 1892 to produce the hammer and several other local inventions.[30] Edward H. Blossom was granted patent number 642,070 on January 30, 1900, for his improvements to the Dupont hammer. He contributed a new method of securing the guides and an effective brake, which consisted of a split ring set into the rear of the drive pulley. In 1902, the hammer was acquired by the Fairbanks Company.[31] Dominating local industry in St. Johnsbury, it was probably far better equipped to build and distribute what then became known as the Fairbanks Power Hammer. Meanwhile, Philippe Dupont, who had not been an officer in the company that bore his name, took a new partner and

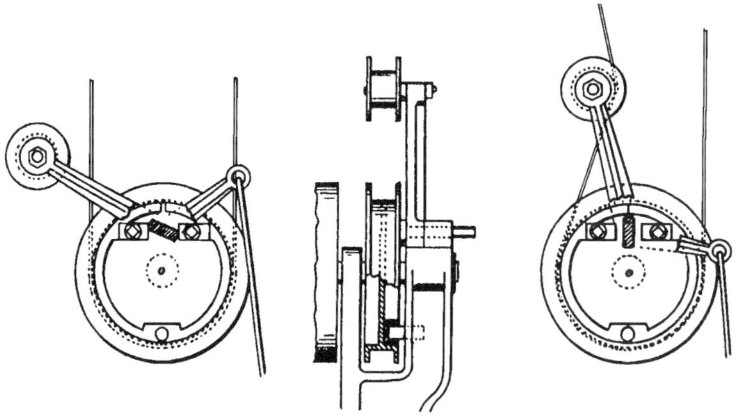

Edward Blossom was granted a patent for his brake design for the Dupont Hammer. When the tensioner pulley was released, a C-shaped steel spring was expanded to contact the drive pulley.

started the Dupont-Anderson Forging Company, specializing in toolsmithing.[32]

An article concerning the Fairbanks hammer, which appeared in the *American Machinist* in 1913, describes the addition of an outboard drive-shaft mounted low at the rear of the machine. This was intended to allow the hammer to be placed in shops which lacked the necessary headroom for a regular hammer model. The hammer was offered in sizes from 25 to 300 pounds, with either belt or motor drive.[33] In 1917, the Fairbanks hammer business was sold to the United Hammer Co., of Boston. At that time, it was said that more than 1,400 of these hammers were in service. An adjustable, tapered bronze gib and redesigned faceplate were offered on new hammers and were also available as a retrofit for earlier machines.[34] The United Hammer Company further enhanced the hammer by offering a protective guard of wire mesh to enclose the moving parts. It also added 400- and 500-pound models to the line. These were intended for heavy forging in railroad repair shops and shipyards.[35] During the depression, the hammer was sold to the Barbour-Stockwell Company of Cambridge, Massachusetts, who also acquired the Beaudry Hammer line. Both of these lines were dropped in the mid-1950's.[36]

* * *

After more than a decade in the power hammer business, the Beaudry Company of Boston, Massachusetts, began to manufacture its Beaudry Champion Power Hammer. This vertically-configured, guided-ram hammer employed a distinctive spring linkage which was patented by Augustin Beaudry (patent number 526,370 granted September 25, 1894, and 545,763 granted September 3, 1895). Beaudry's linkage differed considerably from those previously discussed, and was proclaimed to be "elastic," causing the ram to "rebound the instant the blow is struck without reaction or jump or sudden undue strain on any of the hammer

An early illustration of the Beaudry Champion Power Hammer clearly showing the unique spring-arm linkage. (*Beaudry & Co. trade literature*)

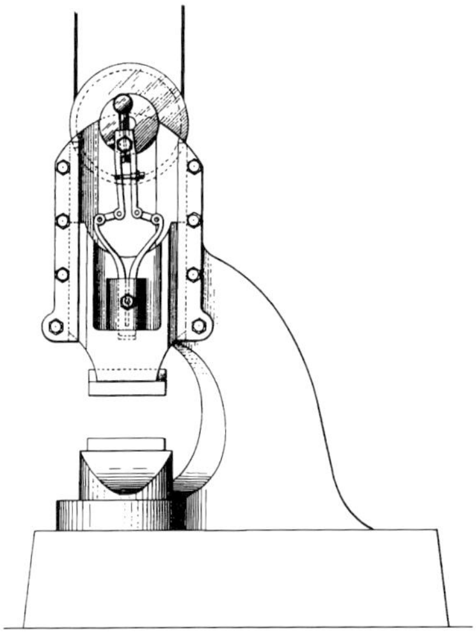

Details from Augustin Beaudry's patents of 1894 (above) and 1895 (below) reveal numerous variations on the theme of Beaudry's unique spring-arm and roller linkage.

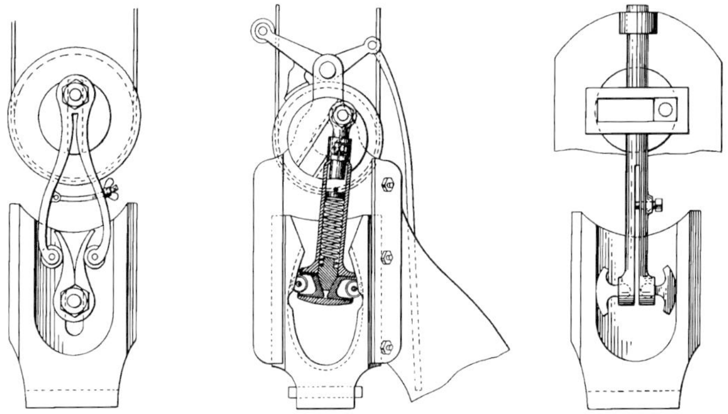

parts."[37] To accomplish this, the ram had an internal, elliptically-shaped track. Two steel spring arms, with hardened steel rollers at their extremities, operated within the ram, upon the curved track, and served to lift and throw the ram. This remarkable approach was extremely successful, and the hammer was built in one of the most complete ranges of any hammer of its type. At one point, eleven sizes, from 50 to 500 pounds were offered, and the company pronounced its line superior for general forging, particularly where there were frequent variations in the thickness of the stock.[38] Like Beaudry's earlier, non-vertically-configured hammers, the frame of the Champion Hammer was distinctively arranged so that the ram and guides were rotated to the side rather than the front. This simple detail (which can also be seen in the earliest Shaw and Justice hammers) gave the smith greater clearance, especially for long workpieces. Of course, some builders simply rotated the dies to accomplish much the same thing, while others provided a hole in the frame, through which a long workpiece could extend.

To complement the Champion Hammer, Beaudry brought out a new design in 1912. The Beaudry Peerless Power Hammer was of a more conventional design in its linkage, while retaining the signature, side-oriented ram. The linkage relied upon the

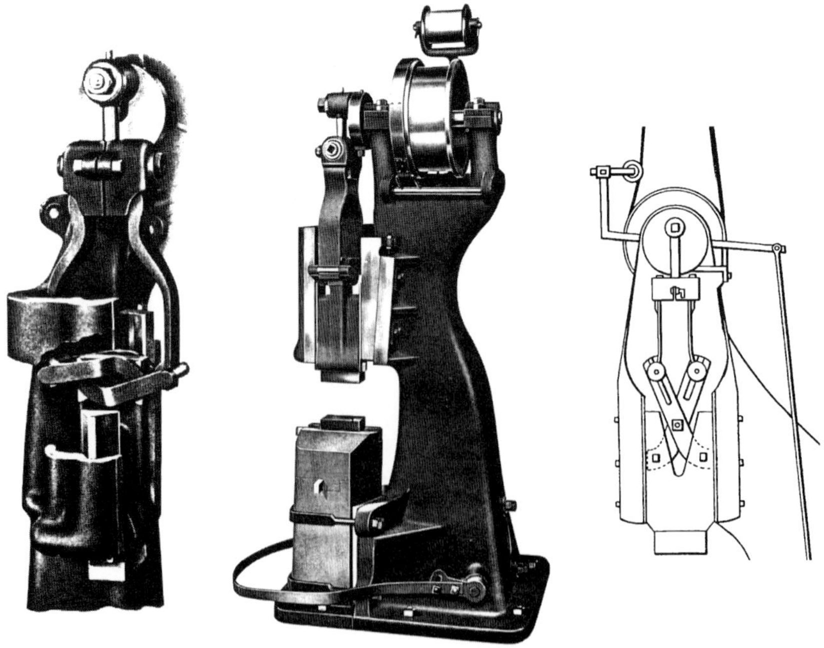

The Beaudry Peerless Power Hammer appeared on the market in 1912. The spring arms were outside of the ram, but the links were unusual in their cross-over arrangement. This link configuration was patented by William Botting of Fitchburg, Massachusetts, who was granted patent number 550,581 on December 3, 1895. (*Beaudry & Co. trade literature*)

flexibility of two spring arms (forged of Swedish steel, and carefully tempered). The ram was attached to the spring-arms externally by links that crossed over one another. The Peerless was offered in seven sizes, from 25 to 200 pounds, and was designed specifically with plating and drawing in mind. Having a quicker action than the Champion, Beaudry recommended it "to manufacturers of cutlery and edge tools, to file, shovel, and agricultural implement makers, and to spindle and general tool makers, as the most practical hammer ever brought out for their line of work."[39] By the early twenties, this hammer had been discontinued. In 1924, a third style, the Beaudry Utility Power Hammer, was introduced. This machine was brought out as a low-cost hammer, in three sizes: 25, 50 and 100 pounds. The still popular Champion Hammer line was reduced at this time to the six heavier sizes, from 100 to 500

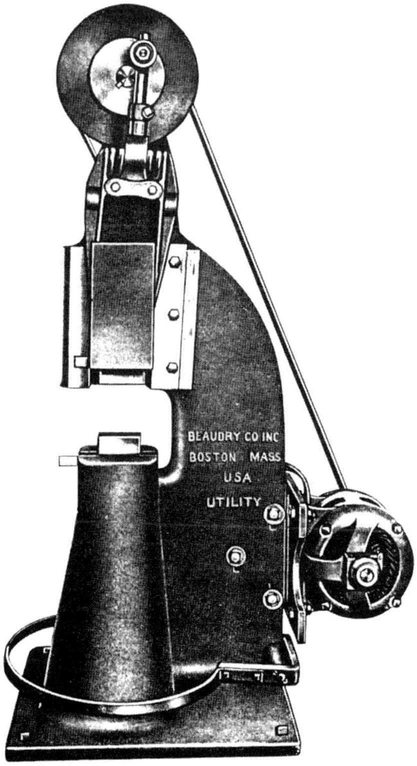

The Beaudry Utility Hammer was placed on the market in 1924 as a low-cost hammer. (*Beaudry & Co. catalogue*)

pounds. The linkage of the Utility Hammer was an interesting variation on that of the Champion. Non-flexing steel arms terminated in hardened steel rollers which followed curved tracks on the *exterior* of the ram. A short coil spring, above the arm pivots, held the arms against the ram.[40] The Utility Hammer was not Beaudry's only new addition at this time. A curious, anachronistic, air-cushioned helve hammer was developed;[41] and in 1925, the company offered a self-contained air hammer.[42] The self-contained air hammer was subsequently bought by the Erie Foundry Company to complement its line of steam hammers.[43] When Beaudry was acquired by the Barbour-Stockwell Company, only the Champion hammer was continued, and in only four sizes (100, 150, 300, and 500 pounds). The Beaudry Company apparently failed shortly before the Great Depression. Several weeks before Black Tuesday, it was announced that Otto Abrahamsen, for twenty years treasurer and director of sales for Beaudry, would be in charge of sales for the Moloch Foundry & Machine Company of Kaukauna, Wisconsin.[44] That company will appear later in this chapter.

* * *

Having failed to acquire the patent rights to the Dupont hammer in 1891, C. C. Bradley & Son, of Syracuse, New York, was slow to enter the vertically-configured, crank-actuated guided-ram hammer market. In 1899, however, the Bradley Compact Hammer was added to its extensive line of helves and helve-like, guided-ram machines. The company emphasized its suitability where floor space was at a premium, and a high-speed hammer was desired.[45] Weighing somewhat less overall than Bradley helve hammers of similar capacity, the Compact was also somewhat more affordable. A 50-pound Compact with electric motor drive cost $700 in 1939, compared with $850 for a similarly equipped Upright Strap Hammer. The true economy of the Bradley Compact Hammer becomes debatable when one considers the $292 price tag on a 50-pound Little Giant.[46] The Bradley Compact Hammer was, however, extremely robust, and built with the industrial user, rather

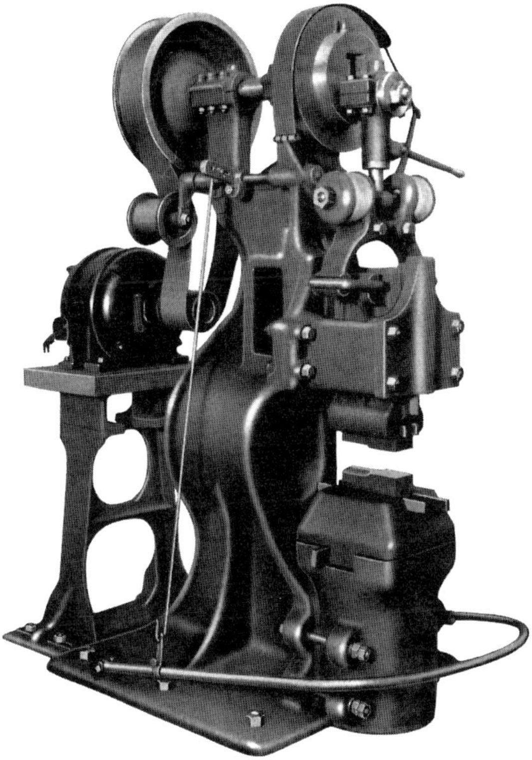

The Bradley Compact Hammer with electric motor drive. Note the rubber-cushioned spring-yoke linkage, and the lever which allowed the pitman to be readily adjusted. (*Bradley Co. catalogue*)

than the rural blacksmith, in mind. The Compact used a spring-yoke arrangement that was similar to the one used by Williams, White. Two of Bradley's hallmark rubber cushions were horizontally-oriented to compress the arms of the yoke. The arms were attached to the ram by thick, rectangular steel links. This hammer was initially offered in sizes from 15 to 200 pounds. Eventually, a 300-pound model was added, and nine sizes were offered through the Second World War. The 15-, 50-, 75- and 125-pound models were discontinued shortly after the war, with the remaining 30-, 100-, 150-, 200- and 300-pound hammers soon to follow.

* * *

Another Syracuse manufacturer began building vertically-configured hammers in about 1902, giving the Bradley Company competition in its own backyard. By 1919, the Kane and Roach Up-To-Date Power Hammer was being built with a linkage which was similar to that used on the Bradley Compact Hammer. The Kane and Roach machine employed steel coil springs (unlike Bradley's rubber cushions) in precisely the manner described in Leonard D. Howard's 1894 patent. It is not known if Kane and Roach licensed the use of this design, or simply adopted it after the patent had expired. If the latter is true, the earliest hammers probably used a different linkage design. It was claimed that William E. Kane designed every machine in the company's catalogue, including the hammer, but there were no Kane patents associated with this product.

The company also claimed to "have discovered practically all of the weak and objectionable points in hammers," which suggests that its early models left something to be desired. However, by 1919, there was only pride in "having eradicated every weak and objectionable point and... reached the nearest point to perfection and convenience of any hammer on the market."[47] The Up-To-Date Power Hammer was offered in five sizes : 35, 50, 85, 125 and 150 pounds. All sizes possessed a hole in the frame behind the dies, which allowed long stock to be worked along the length of the dies. This was a fairly common feature, but the Up-To-Date Hammer also boasted a solid steel anvil (available as an option on only a few other hammers). Another rare feature of the Kane and Roach hammer was the placement of a guard around the crank and linkage. This guard served a dual purpose: protecting the smith (though incompletely) in the event of breakage, and functioning as a rest for welding "low" (small-diameter) tires. This rest was complemented by special slanting dies which compensated for the outward tilt of the tire. On the 50-pound hammer, for example, the slanting dies were required for tires less than 34 inches in diameter. Larger-diameter tires would extend over the guard, and could be welded with standard dies. The company guaranteed that its hammer would "pay for itself clear at least three times a year, and earn you more than any other three machines in your plant."

Kane and Roach was well-established by the time it entered the hammer market. The enterprise was founded originally by William E. Kane in 1887 as the Monarch and Eureka Company. Its first line consisted of chemical boiler filters. Bartholomew F. Roach became a partner in the firm in 1891, and contributed substantially to the business side of the company until his death in 1900. The true strength of the company, though, was the inventive genius of William Kane. As the company expanded into the

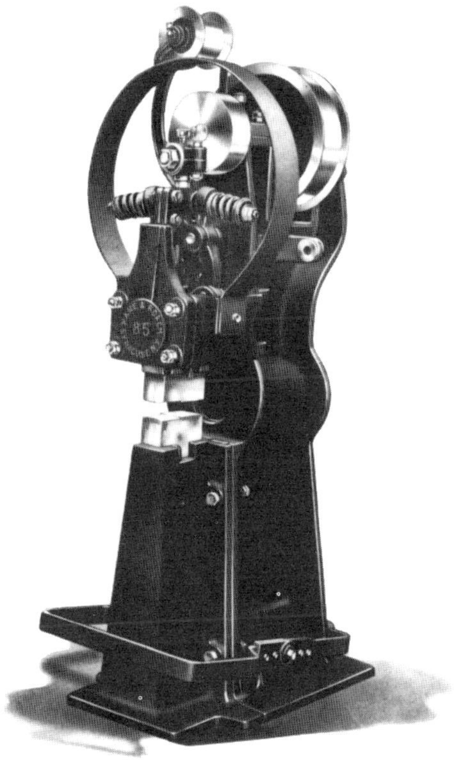

An 85-pound Kane & Roach Up-To-Date Power Hammer. The circular guard also served as a rest for supporting small-diameter tires when welding. (*Machinery*)

carriage and wagon machinery line, Kane contributed a power hub borer. In the 1890's, Kane developed his first straightener for use in steel mills. With the straightener, cold roll forming machines and a wrapping type bender, the company rose to world-wide prominence. In 1937, with the power hammer still in production, Kane and Roach was building over 70 sizes and types of straightening rolls and over 60 types of forming machines. This variety notwithstanding, the majority of its actual production in this line was custom work. The firm was undeniably successful in its desire to be "a small plant dedicated to quality manufacture."[48] It is curious that Kane and Roach should enter the fiercely-competitive power hammer market, but its commitment to this line was solid. The company was listed in the *Thomas' Register of American Manufacturers* through the Second World War.[49]

* * *

While metallic and rubber spring linkages quickly took the lead in crank-actuated hammer design, the air-cylinder linkage remained in production. In the late 1880's, a new atmospheric hammer appeared in America, offering competition to Hotchkiss' durable design. The vertically-configured Hackney Hammer was actually an import, an English design based on the earlier American innovation. It was developed by Gilbert Glossop, of Sheffield, and was patented in England on February 3, 1887 (number 1,706). On December 11, 1888, two American patents were granted to Glossop, but assigned to Fredrick Clement Brooksbank and Jacob B. Perkins, both of Cleveland, Ohio. The first of these, patent number 394,483 pertained to a friction brake, while the second, patent number 394,484 described the distinctive valve arrangement used on the cylinder of Glossop's hammer.

The air-cylinder of Hotchkiss' hammer employed only one "valve," which was simply a hole located at the mid-point of the cylinder's length. This enabled the air-cylinder to act as a spring in both directions, to store energy for an enhanced blow and to cushion the mechanism from shock. The hammer ram was placed

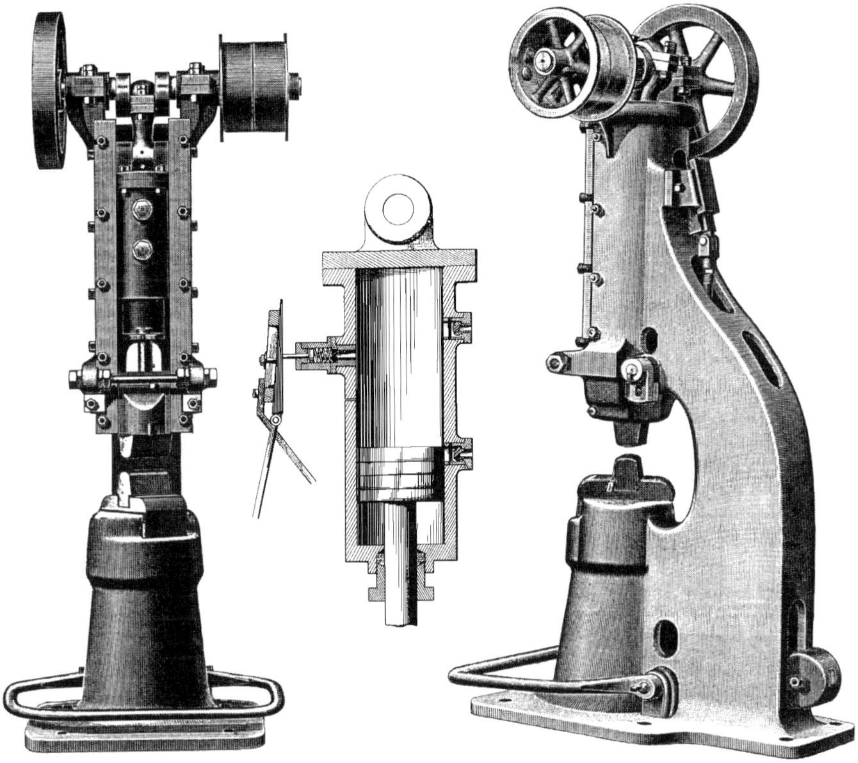

The earliest Hackney Pneumatic Power Hammers sold in America were actually English imports. (*American Machinist*) Gilbert Glossop's patent of 1888 (#394,484) described his valve arrangement.

in motion whenever the cylinder moved.[50] The Glossop hammer was quite different. An adjustable exhaust valve in the upper half of the cylinder, combined with one-way inlet valves in the upper and lower halves, enabled the stroke to be not only cushioned, but controlled. With the exhaust valve open, the air-cylinder moved up and down without disturbing the piston and ram (conceptually, not unlike the Welch and Blacker crank-actuated helves). As the exhaust valve was tightened (by movement of a sliding wedge) the air-cylinder began to affect the ram. A rapid closure of the exhaust valve produced an immediate hard blow. Similarly, the hammer stopped immediately when the valve was released. The friction brake served to arrest the motion of the ram in its upper position.

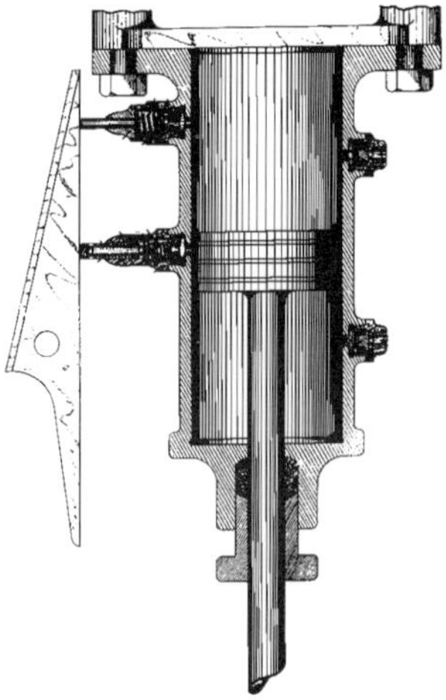

Gilbert Glossop's patent of 1892 described the addition of a second valve, which increased control of the hammer's stroke.

Hammers with Glossop's variable air-cylinder spring resemble the self-contained pneumatic hammers which rose to prominence in the twentieth century. However, these later hammers were truly air-actuated, using a secondary, compressor cylinder to actuate the ram. It is probably no coincidence that the earliest American version of the self-contained pneumatic hammer was developed in Cleveland (by Melvin Yeakley in the mid-1890's).[51]

On November 1, 1892 Gilbert Glossop was granted an additional patent (number 485,498) which described a modification of his air-cylinder valves. Instead of one adjustable exhaust valve controlled by the sliding wedge, two were proposed: one at the mid-point and another closer to the top of the cylinder. This arrangement provided still more control, enabling the hammer to deliver blows of varying force at different heights above the anvil. In the case of small hammers, where little variation in stroke was required, Glossop recommended an open port in lieu of the upper variable exhaust valve. Slight variations could be obtained by regulating the cushion of air below the piston with the middle exhaust valve.[52] On the same date that Glossop was granted his patent for this improvement, Fredrick Brooksbank was granted two patents, in his own right, for hammer inventions. The first of these,

An atmospheric forging hammer of Glossop's design (possibly an American-built Hackney hammer) at work in the forge of Carl Flohr in Berlin. (*Carl Flohr - Fest-schrift zur feier des 25 jahrigen Geschsftsjubilaums - 1904*)

patent number 485,168 described a curious variation applicable to either crank-actuated atmospheric hammers or steam hammers. In order to use several dies in one hammer at the same time, Brooksbank proposed building a hammer with two cylinders which were rigidly connected. The two piston rods would thus support a broader work area, diminishing the likelihood of bending or breaking the rods. Hammers of this design were probably never manufactured. Brooksbank's second patent, number 485,169 describes a modification of Glossop's friction brake, which included a means for easily renewing the wearing surface of the brake.

The Hackney Pneumatic Power Hammer first appeared in the pages of *American Machinist* in September of 1888. At that time it was already being made in seven sizes : 50, 75, 100,150, 200, 300 and 500 pounds. Since interested parties were referred to selling agents Hill, Clark and Company, of Boston, it is likely that these were imported Glossop's hammers, built in England by

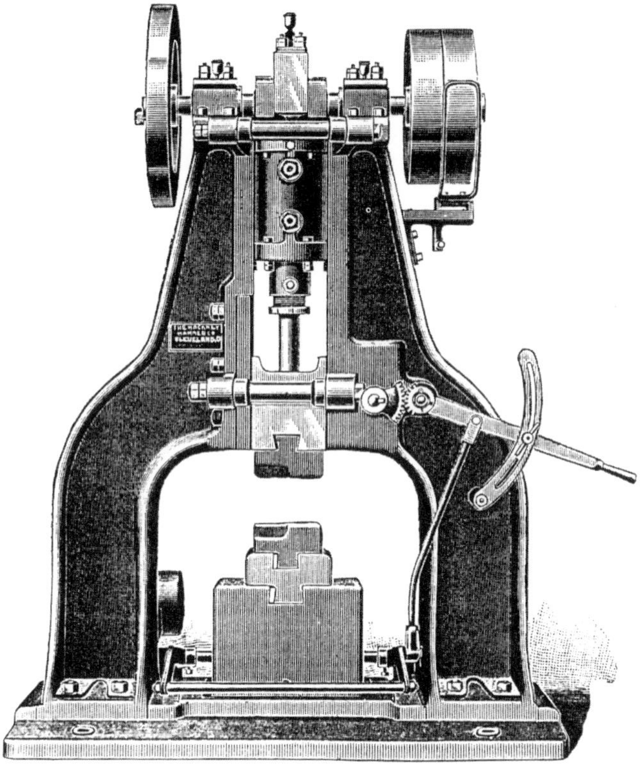

A 300-pound Hackney double-frame hammer. One of the few crank-actuated machines built in this configuration (*American Machinist*).

H. Broadbent of Sowerby Bridge.[53] Earlier in the year, a 5 cwt. double-frame hammer of Glossop's design had appeared in *The Engineer*, "specially designed...for steel tilting and forging, plating hoes, spades and shovels, scythe skelping, etc."[54] In 1889, the Hackney Hammer Company was formed in Cleveland, Ohio, with Jacob Perkins as president. Fredrick Clement Brooksbank was vice president, having previously acted as an agent for British steel manufacturer James Jackson, of Sheffield.[55]

The modern shop methods of the new Hackney Hammer Company attracted the attention of an editorial correspondent for *American Machinist* in April of 1891. The company was pronounced to be "one of the best examples of the proper way to start in to the manufacture of a machine..." Its shop was not a

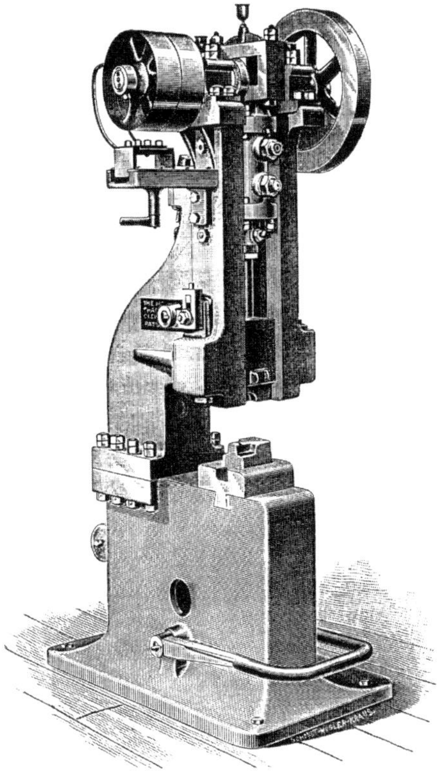

A 50-pound Hackney hammer. (*American Machinist*)

comprehensive manufacturing facility, but rather an erecting shop which received castings and forgings from third parties. The erecting floor was dominated by a single, massive machine tool: a milling and boring machine built by the Niles Tool Works. Upon this machine, with its nine feet of horizontal travel and six feet of vertical travel, holes were drilled and surfaces milled. State-of-the-art milling cutters (some up to sixteen inches in diameter), with removable inserted teeth, were used. The correspondent observed that "the hammer seems to be a case in which the milling operation can be applied to the very best advantage, and excellent use has been made of the opportunity."[56] The Hackney hammer itself was again profiled by *American Machinist* a month after this editorial. Engravings of a 50-pound single-frame hammer and a 300-pound double-frame hammer were presented, though only six sizes were

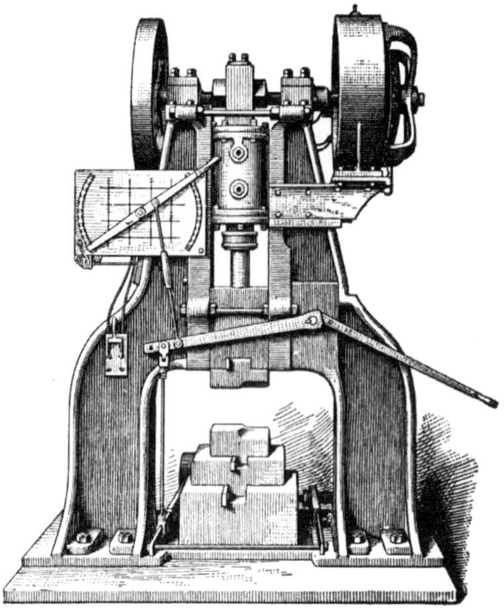

Direct drive of a machine by an electric motor was rare in the 1890's. Here is a Hackney hammer with a shockingly simple speed control. (*The Iron Age*)

apparently in production at the time.[57] *Iron Age* followed suit with a description of the Hackney hammer in June of 1891. It was noted that, "the hammer head stops instantly upon withdrawing pressure from the treadle, so that even drop forging has been done satisfactorily under the hammer without any special preparations."[58] In May of 1892, *The Iron Age* offered a description of a Hackney Double-Standard (double-frame) Power Hammer. This larger size of the Hackney hammer included Brooksbank's improved brake.[59]

A catalogue issued by machinery dealer E.A. Kinsey, of Cincinnati, Ohio (from about 1893), included the full extent of the Hackney Hammer line: there were seven sizes, with 50-, 75-, 100-, 150- and 200-pound hammers available with single-frame, and 100-, 200-, 300-, and 500-pound hammers available with double-frame construction. The smallest machine weighed 3,000 pounds overall, stood six feet two inches tall, and struck 380 blows per minute. The 500-pound, double frame hammer weighed seven tons overall, stood eight feet five inches tall, and struck 225 blows per minute.[60] A portion of the Hackney hammer line was carried by Manning, Maxwell & Moore in 1894. The 75-pound hammer was absent, and only the 300- and 500-pound hammers were offered with a double-frame. It was noted that hammers up to and including the 150-pound model were built in two parts. Since the anvil was integral to the lower part, the entire lower section could be replaced "with little trouble and expense" should this part be broken. Hammers of 200-pounds and up were wisely built with separate anvils. A 100-pound Hackney hammer cost $600, which was comparable in price to a vertically-configured machine of Belden's improved design. At the same price one could also buy a 100-pound Bradley Upright Strap Hammer. The 500-pound Hackney hammer cost $1,200, which placed it into competition with the numerous steam hammers available at the time.[61] Competition from air-actuated, self-contained pneumatic hammers had not yet appeared.

Despite favorable initial publicity, reasonable pricing, and representation of the line by the foremost of American machinery suppliers, the Hackney Power Hammer was short-lived. Even as production of the hammer was beginning, its builders diversified. The Hackney Bicycle Manufacturing Company was formed to take advantage of the bicycle craze that was sweeping America.[62] The League of American Wheelmen had been formed several years earlier, and the campaign for better roads was in full swing. At first, the bicycle works was located in the same shop on Johnson Street. It was soon relocated; and by the time the 1894 *Cleveland City Directory* was compiled, the Hackney Hammer Company no longer existed.[63] It is likely that the builders of the Hackney hammer had been seduced by a more profitable manufacturing enterprise.

* * *

While many of these diverse, vertically-configured, crank-actuated hammers were intended primarily for heavy industrial use, an increasing number of the lighter-weight hammers were finding application in small shops. The availability of cheap power, combined with high labor costs, and a healthy demand for blacksmithing services, made power hammers increasingly attractive to rural smiths. Many small foundries were quick to appreciate this fact, and the first two decades of the twentieth century saw a remarkable boom in small, low-cost power hammers. Dozens of companies entered the market, advertising directly in magazines such as *The Blacksmith & Wheelwright* and *American Machinist.* Both guided-ram and helve hammers were built in great numbers. Many of these machines were well-designed and protected by patented features, but there were also those which simply copied the more successful models. Ultimately, the most successful crank-actuated hammer of them all was the Little Giant.

More than 18,000 Little Giant Power Hammers were built in Mankato, Minnesota, from 1896 to 1984.[64] So completely did it dominate the industry, that in later years it was claimed that the founders of the company, the Mayer Brothers, had actually invented the power hammer![65] This myth was promoted by the company itself. In 1926, for example, the Little Giant Company proclaimed its golden anniversary (somewhat prematurely!) It harkened back to the patriotic, Centennial year of 1876 as the year in which a village smith in Mankato built his own "trip hammer," thereby launching a great industry![66] It is quite possible that German-born blacksmith Lorenz Mayer did build his own hammer - many smiths did - but the commercial production of hammers in Mankato did not begin until the mid-1890's; and then, in the machine shop and foundry established by Lorenz's three sons, Louis, Lorenz, Jr. and Conrad (Conrad left the firm at an early point). While the Mayer Brothers' hammer did initially possess a novel linkage, patented by Louis in 1899, the hammer was built along well-established lines. Also, after only a few years of production, a less-original but more durable and effective linkage was adopted. For almost a decade, the Mayer Brothers limited their hammer production to one fifty-pound size. Only later did they fully enter the industrial hammer market which was dominated, at the turn of the century, by Beaudry, Fairbanks, Bradley, and others. The story of the Mayer Brothers and the Little Giant is complicated by the fact that Louis and Lorenz

Mayer, Jr. remained in the hammer-building business long after they were voted-out of the company which bore their name. The brothers moved first to Antigo, Wisconsin, and later to Kaukauna, manufacturing a hammer that was very similar to the Little Giant, which meanwhile continued to be built in Mankato. The later Mayer Brothers' hammer was eventually acquired by the D.J. Murray Company, of Wausau, Wisconsin. Their Murco Power Hammer was ultimately bought by the Little Giant Company in the mid-1960's: a remarkable half-century circle![67]

The Mayer Brothers did not initially set out to build hammers. In 1891, they established a general machine shop, adding a foundry in 1894. They advertised a wide range of capability, noting a specialty in jail cells and prison work. Their first thirty hammers were sold in 1896; and, in April of 1899, they began to advertise the New Little Giant Trip Hammer in *The Blacksmith & Wheelwright.* Claiming over seventy hammers in use, they offered it as, "the best and cheapest hammer on the market today for plow-shops and general forge work."[68] This hammer featured Louis Mayer's "yielding ball-coupling," for which he had been granted U.S. patent number 619,048 on February 7, 1899. This linkage was no doubt inspired by the one used on Beaudry hammers. However,

This early advertisement shows the "yielding ball-coupling" linkage patented by Louis Mayer in 1899. (*The Blacksmith & Wheelwright*)

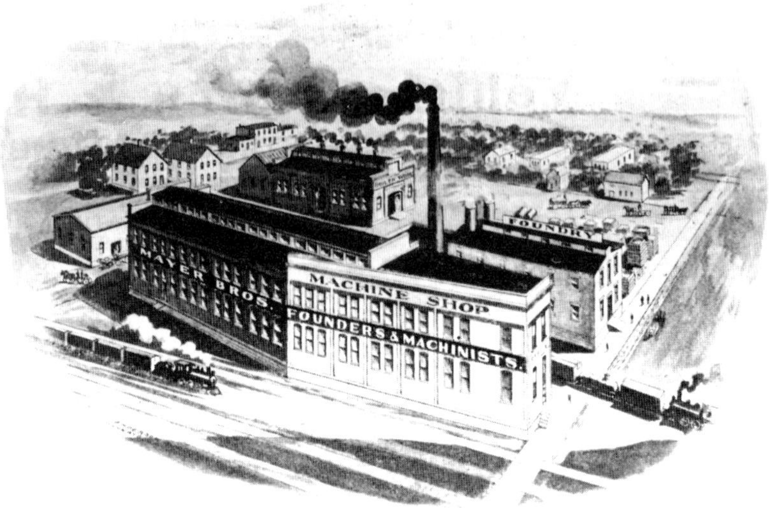

The Mayer Brothers' factory from an advertisement of 1903. (*The Blacksmith & Wheelwright*). Louis Mayer's patents of 1899 and 1901: on the left, his yielding ball-coupling ram, on the right, a Dupont-style spring linkage with simplified connecting links. The ends of the links which pivot in the ram are T-shaped, captured in place by a cover-plate.

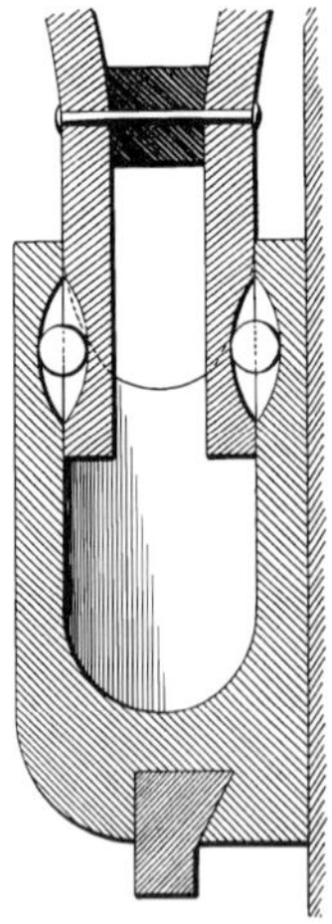

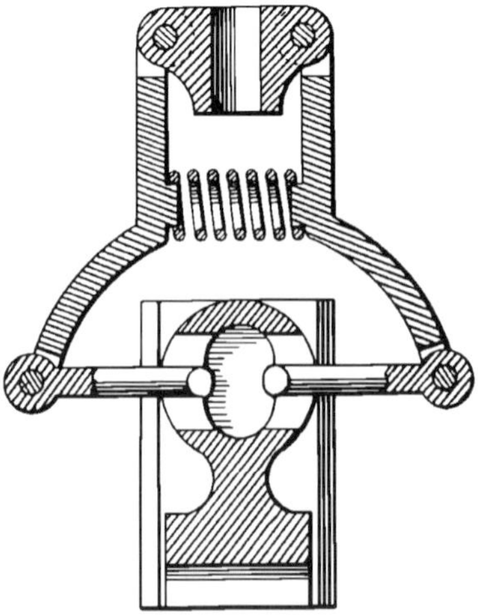

instead of using rollers at the ends of the spring arms, Mayer patented a variation which captured steel balls in pockets formed by recesses in both the ram and the "coupling arms." The arms were held apart by either a rubber cushion or a coil spring. While this arrangement was evidently viable, Louis Mayer continued to contemplate the challenge of a durable, easily-produced linkage, one which would isolate the machine from the shock of impact. Toward that end, he was granted a second patent, number 665,249 on January 1, 1901. This design featured essentially the same toggle arm and spring arrangement developed by Dupont in 1890, but employed a new, simple connection between the toggle arms and the ram. This revised linkage appeared in production at some point in 1900. With the Dupont patent still in force, it is surprising that the new design was allowed to go unchallenged. Like the earlier "yielding ball-coupling" version, the revised hammer held its ram in guides oriented front-to-back. In 1930, a dove-tail slide arrangement, with side-to-side orientation of the guides, was adopted.

The power hammer market was complex at the turn of the century, with many smiths favoring the hammering action and accessibility of the helve hammer. For jobs such as tire-welding, it was hard to beat the unobstructed clearance provided by a helve; and if floor space was not a major consideration, there were a great many helve hammers from which to choose. Appreciating this, the Mayer Brothers brought out their own version in 1902, which they dubbed the "Easy" (discussed in the previous chapter). Another unusual hammer which was produced alongside the popular Little Giant, was a modified version intended for "end-ironing neckyokes, singletrees and similar work." In lieu of an anvil block and lower die, this specialized hammer offered a series of shelves which could support a plate at the appropriate distance for a particular job. The Neckyoke Singletree Hammer was a dedicated "upsetting" tool which answered one of the main shortcomings of almost all open-die power forging hammers. It disappeared from company trade literature at about the same time as the Easy.[69]

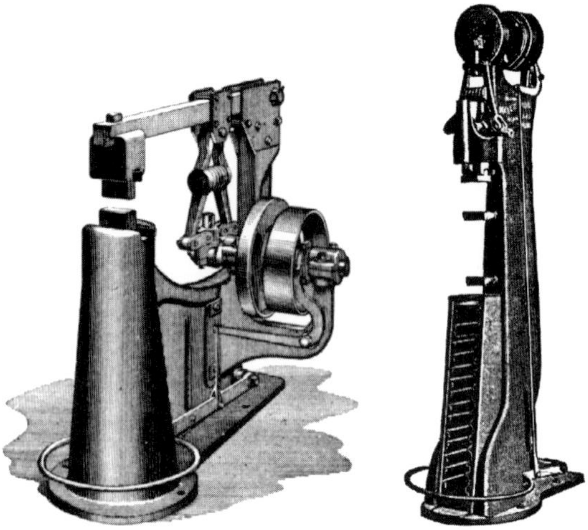

In addition to the Little Giant, the Mayer Brothers built the Easy Helve and the Neckyoke Singletree Hammer. The latter machine was a dedicated upsetting tool: instead of a solid anvil there were brackets for a movable foot-plate. (*The Blacksmith & Wheelwright*)

With the 50-pound Little Giant hammer well established in a growing market, the Mayer Brothers brought out a 100-pound model in 1905, and added a 25-pound hammer in 1908. The company incorporated in 1903 as Mayer Bros., Inc., and pursued a wide range of products besides its hammers. Louis developed and built a prototype V-8 engine, which was installed in a Mayer-built chassis. Other vehicles, including a four-wheel drive truck, fire engine, road grader, and tractor attracted the brothers' attention and energy. Despite the continuing success of the power hammer, Mayer Bros., Inc., failed. After re-structuring, the firm emerged as the Mayer Bros. Company. On-going difficulties in the manufacture and marketing of the tractor ultimately damaged the brothers' credibility, and the Board of Directors voted them out at the end of 1915. By the fall of the next year, Louis and Lorenz, Jr. were established in Antigo, Wisconsin, building the Mayer Power Hammer. Until the beginning of 1919, blacksmiths were likely perplexed to find advertisements (sometimes on the same page) for both the Mayer Brothers and the Mayer Brothers Company. The latter began to emphasize the Little Giant name in 1918, and soon became the Little Giant Company. The war years heightened an already vigorous market for power forging hammers. The company began production of the 250-pound hammer in 1916, and built its first 500-pound hammers soon thereafter. Trade literature from the early 20's includes a 1,000-pound model which weighed almost ten tons overall and was priced at $2,800 (compared to $210 for the 50-pounder). It is unlikely that many of these not-so-little giants were built.[70]

The Mayer Brothers left Antigo after less than three years, forced-out by the sale of the building which they had leased.[71] By September of 1919, a full line of Mayer Power Hammers, in five sizes, was being offered by the Kaukauna Machine Works, in

The Little Giant Power Hammer continued to be built in Mankato after the Mayer Brothers left for Antigo, Wisconsin. (*The Blacksmith & Wheelwright*).

Kaukauna, Wisconsin. *The American Machinist* reported in January of 1920 that the company had redesigned its line of Mayer Power Hammers.[72] Later in that year, the Kaukauna Machine Works merged with the Moloch Stoker Company, of Chicago. The resulting Moloch Company continued to build the Mayer hammer, which was renamed the Moloch Power Hammer. The name "Moloch" refers to an Old-Testament god of fire and sacrifice. While it might seem admirably suited to a machine designed for beating upon fiery-hot steel, the name originated in reference to a far more voracious and demonic coal stoker.

In the years following the brothers' departure from Mankato, Louis continued to refine and patent new power hammer features. On May 8, 1917, he was granted patent number 1,225,046 which protected a new design for the links which connected the toggle arms to the ram. Instead of T-shaped ends, the ends were merely hooked. This simplified the manufacture of the ram, and made it possible for the links to be repaired with minimal expense and expertise. This feature is characteristic of all of the post-Mankato, Mayer Brothers' hammers. Another Mayer patent was granted on August 12, 1919 encompassing improvements for guides of a curious, non-vertically configured hammer, "more particularly of

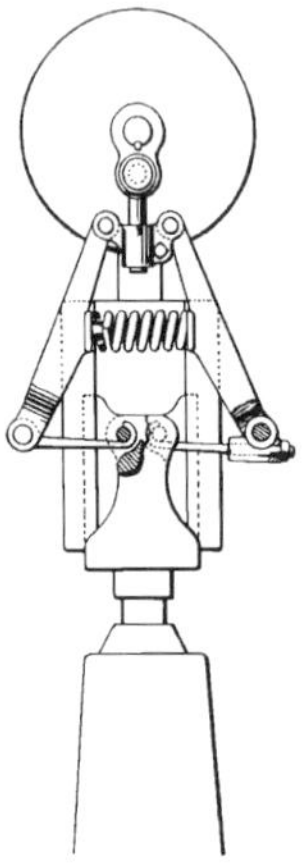

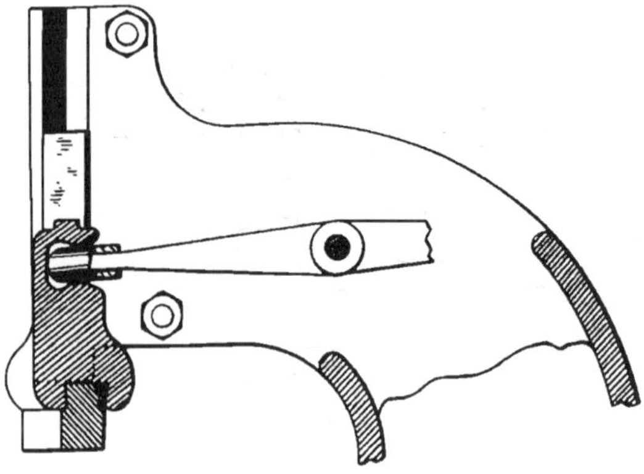

After leaving Mankato, Louis Mayer continued to patent his hammer developments. In 1917, he patented a further simplification of the toggle-arm links (now adjustable). In 1919, he patented an unusual guided-ram hammer variation.

that type including a reciprocal ram carried in the guide head..." It would be interesting to know if this design was developed for another manufacturer (perhaps David Joy, who designed hammers of this type in England), or was patented simply with a future Mayer Brothers' project in mind.[73] In yet another patent from the Kaukauna period, Louis directed his attention to simplifying the operation of cutting-off metal stock under a power hammer. Patent number 1,386,891 (granted August 9, 1921) described a cleverly-configured lower die-block that included a portion dedicated to cutting. A four-sided cutting tool could be rotated as the upward edge became dull. Custom dies were an important part of power hammer sales, with Little Giant offering special "pick point & chisel bit dies" that included a simple cutting feature.[74] It is not known if Louis' rotating cutter was ever brought to market.

Eventually, the Mayer Brothers and the Moloch Company went their separate ways. In early 1929, Louis, Lorenz, Jr. and Bertram Mayer filed articles of incorporation to become, again, the Mayer Brothers Company. They had purchased the old Crawford & McKrimmon Foundry & Machine Shop in Brazil, Indiana, and "planned to begin manufacture on a large scale of mining machinery, pumps, power hammers, boilers and engines." Unfortunately, 1929 proved an inauspicious year in which to commence an industrial venture, and it is unlikely that power forging hammers were ever built in Brazil.[75]

The Moloch Company foundered in the years following the Great Depression. In 1939, the Moloch hammer business was purchased from its receiver by the D.J. Murray Manufacturing Company of Wausau, Wisconsin.[76] This large, prosperous concern had been in operation for more than half a century. Business centered initially on building and repairing sawmill machinery, but ultimately shifted to pulp and paper equipment. These lines required the establishment of a large foundry. Concerns over the economics of the timber industry in Wisconsin led to thoughts of diversification; and with America in preparation for the second World War, the Moloch Power Hammer was undoubtedly a smart acquisition. The machine naturally became the Murray Power

For more than two decades, the chief competitor of the Little Giant was the Murray, later Murco, Power Hammer. It was built in five sizes from 25 to 500 pounds. (*D.J. Murray Co. trade literature*)

Hammer, (the word "improved" appeared in some advertisements, though the hammer was little-changed from its earlier, Moloch days). The "Murco" trade name was coined in the early 1940's, and the hammer took its place in a line that included another realm of diversification: hoisting machinery. Eventually, in the mid-1960's, the Murco Power Hammer was sold to the Little Giant Company. This acquisition was probably motivated by a desire on the part of Little Giant to eliminate competition in an already dwindling market. The Little Giant Company certainly did not need the designs, patterns and name, but if they did not buy the Murco Hammer, someone else might. The D.J. Murray Manufacturing Company became Murray Machinery, Inc., in 1966. The plant closed in 1986 and was demolished.[77]

Following a second bankruptcy in 1937, the stock in the Little Giant Company was bought-out by Leonard Fazendin, who had run the company since 1923. The company prospered during the years of the Second World War, selling almost three thousand hammers - an important contribution to the vital wartime industrial effort. After the 1940's though, hammer sales plummeted, and the company turned to its foundry to remain in business. In the hands of Fazendin's son-in-law, Henry Dotson, the foundry was modernized and prospered. Ultimately, the foundry business became the Dotson Company, a partnership between Dotson and Fazendin, and was run separately from the hammer-building Little Giant Company. After Fazendin's death in 1955, the companies merged under the Dotson Company name. Over the next thirty years, only about 500 hammers were built. The last Little Giant hammer, a 50-pounder (serial number 7263), was built in 1984. In 1991, the Little Giant Company (name, parts, patterns and records) was sold to Harlan Suedmeier, of Nebraska City, Nebraska. "Sid" Suedmeier has made a wide range of replacement parts available for the thousands of Little Giants that are still in service.[78]

* * *

Thus, as the twentieth century dawned, there were many different vertically-configured crank-actuated hammers on the market. The majority were costly machines intended for manufacturing rather than general blacksmithing. In America, the blacksmiths' profession was flourishing, as yet unaffected by the emerging age of the automobile. With newly-installed gasoline engines, rural blacksmiths could realistically consider the advantages of "a helper with no bad habits," and many inventors and foundry operators rose to the challenge of supplying an affordable hammer. But the blacksmith is pre-eminently a tool-making craftsman, and many smiths simply built their own. As should be expected, many of the inventors of patented and mass-produced hammers were first-and-foremost practical blacksmiths. Their hammers were often developed initially to meet the needs of their own shop.

The pages of *The Blacksmith and Wheelwright* contain numerous examples and illustrations of the simple power hammers that were built and employed by American smiths in their own shops. Blacksmiths such as A. B. Hallberg, of Canada, wrote asking for "a sketch of a good home-made triphammer with full details of construction."[79] While many of these hammers were trip or crank-actuated helves, the majority were guided-ram hammers of a vertical configuration. A stout post within the smith's shop could

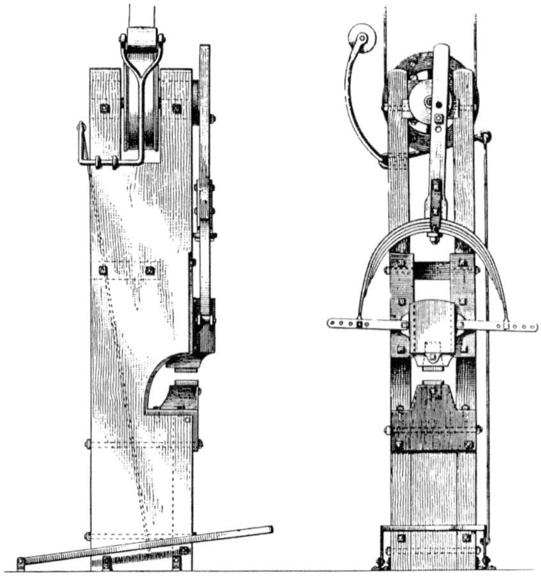

C. E. Rhodes' homemade hammer. (*The Blacksmith & Wheelwright*)

readily become the frame for such a hammer, and every shop was filled with the old springs, U-bolts, steel plates, shafts and pulleys from which a workable mechanism could be fashioned. Usually such a hammer was run profitably until it flew apart. By that time the smith was either wiser, and built himself a better hammer, or better-off and able to afford a brand new Kerrihard or Little Giant.

One such self-built hammer was described in a letter to *The Blacksmith & Wheelwright* by C. E. Rhodes, of Maine. The side-pieces of his hammer were of oak, three inches thick. The crank plate was made from an old mowing machine wheel, the pitman (connecting the crank-pin to the spring) was of wood, and the leaves of the spring were bent to shape without further tempering. The hammer proved durable enough to serve Rhodes for a time and then be sold to another smith.[80] Another variation on this theme was provided by G. W. Allen, of North Carolina. His simple hammer used an existing post for the frame, and a standard anvil.[81] A far more ambitious home-made hammer was that described by H. Feldhaus, of Hallam, Nebraska. Its frame was built from structural-

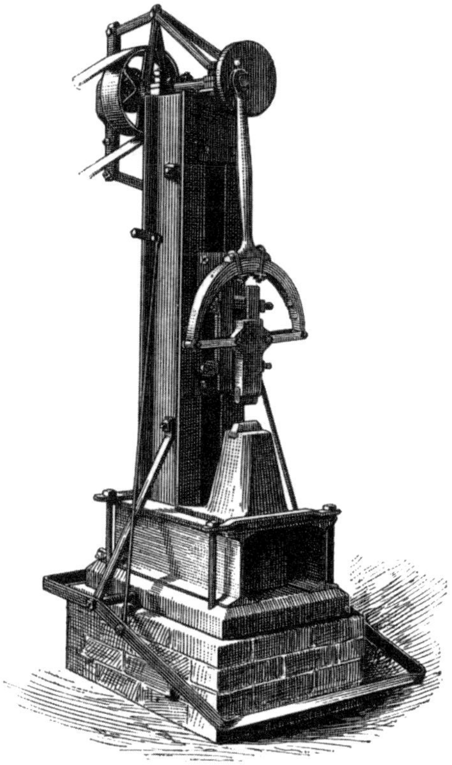

Heinrich Feldhus' vertically-configured Little John Power Hammer began as a homemade machine. (*The Blacksmith and Wheelwright*)

steel I-beams, and the hammer included several cast parts.[82] The response to this design was evidently so strong that shortly after sharing it with his brother smiths, Feldhaus applied for a patent to protect his use of structural steel components. Heinrich Feldhaus was granted patent number 716,019 on December 16, 1902, and his

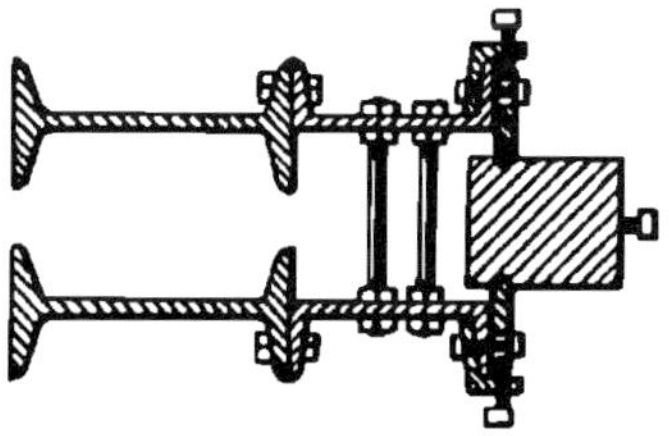

A cross-sectional detail from Feldhus' 1902 patent, showing his use of I-beams for the frame, as well as channel and angle sections for the bracket which supported the hammer ram.

hammer appeared on the market within the year. Dubbed the Little John, Feldhaus intended his hammer to be used primarily for "hammering plowshares, plow-points, and other articles."[83] Several years later, another (probably related) hammer venture emerged from Hallam, Nebraska. The Western Iron & Machine Works asked the smith, "Why not build your own triphammer?" This company offered a kit of castings and machined parts that would presumably save the smith from $25 to $50 if he provided a post or other structure for the hammer's frame. The economics of this offer must not have been pursuasive enough, for the advertisement was short-lived.[84] Another variation on this theme was the Hilbert-Freiberg Power Post Hammer. This forging hammer was configured much like the "bracket hammers" of Dienelt & Eisenhardt. The hammer and drive mechanism were sold as a complete unit, ready to be mounted to a stout post in the smith's shop. The anvil casting was bolted to a wooden block which was, in turn, strapped to the post. Built at the turn of the century by the Hilbert-Freiberg

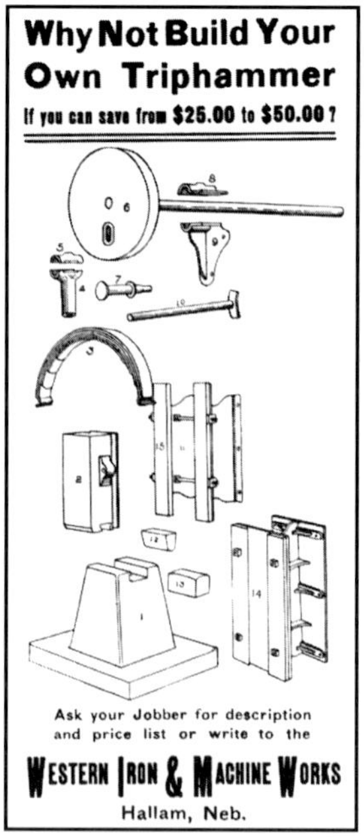

The Western Iron & Machine Works offered a kit hammer. (*The Blacksmith & Wheelwright*)

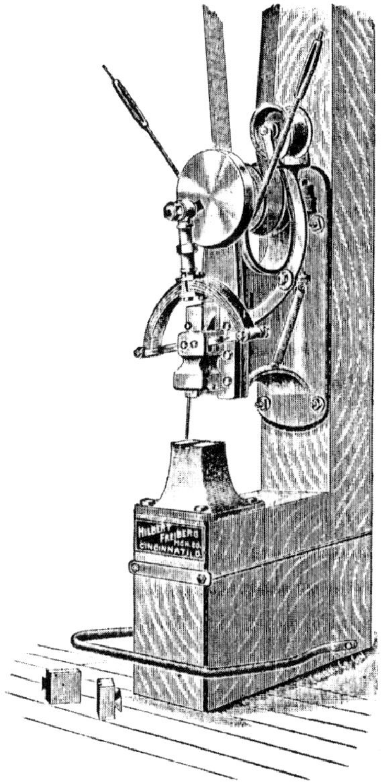

The Hilbert-Freiberg Post Hammer and a similar machine offered by J. A. Kauffman. (both from *The Blacksmith and Wheelwright*)

Machinery Company of Cincinnati, Ohio, this hammer had a 50-pound ram.[85] A similar hammer, which included a post-mounted countershaft, was manufactured by J. A. Kauffman, of Miamisburg, Ohio.[86]

* * *

The McLane Power Hammer was also among those power hammers developed initially for use in the inventor's own shop. This machine ultimately became better-known as the Champion "Patented" Power Hammer, and it was built for many years by the Champion Blower & Forge Company, of Lancaster, Pennsylvania. The salient features of this hammer were patented in 1902 by Jared

Brown McLane, of North Reading, Massachusetts, who was granted patent number 703,737 on July 1, 1902.

McLane was born in Nova Scotia, in 1853, and left school early to apprentice with a blacksmith and carriage builder. He moved to North Reading in the difficult years following the financial panic of 1873. Since there were already two blacksmiths in town, McLane became a jack-of-all-trades. He worked on the public roads, "pathed snow" during the winter, painted signs, and contracted to provide iron railings for local bridges. Ultimately, his forge grew to become a successful wagon-building factory, employing on the average fifty men, including blacksmiths, carpenters, cabinet makers, and sign painters. By 1902, he issued an illustrated catalogue containing all manner of commercial wagons,

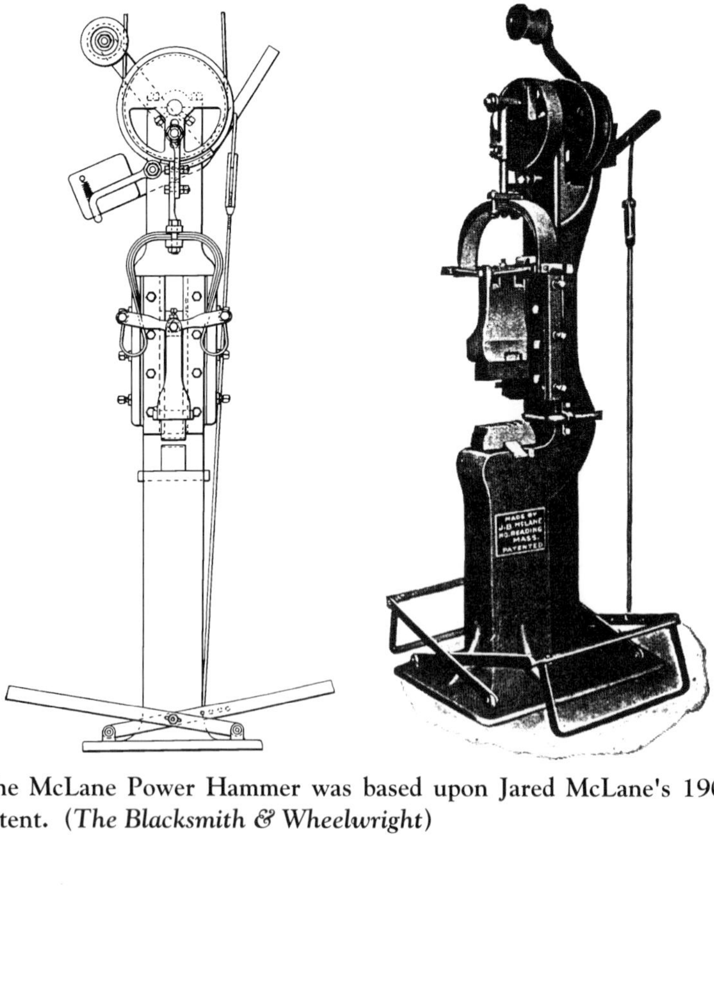

The McLane Power Hammer was based upon Jared McLane's 1902 patent. (*The Blacksmith & Wheelwright*)

Jared B. McLane, builder of "the hammer that gets busy at once and does things!" (*The Blacksmith & Wheelwright*)

"built by our design or your specifications."[87] There was no question of the utility of a good power hammer to this line of work, and McLane's hammer was reportedly born of necessity and experience in his own shop. Though developed at the turn of the century, the McLane hammer was not widely marketed until 1908. Touted as "the hammer that gets busy at once and does things," the hammer featured a distinctive, double foot treadle.[88] This arrangement gave the smith better drive control from all angles, and permitted two smiths to use the machine comfortably between two forge fires. But McLane's patent design was far more comprehensive, including a means of compacting the spring (via a "W" shape, rather than the simple bow). McLane also stabilized the ram in easily adjustable and replaceable guides, and provided for braking the hammer. However, by 1907, the "W" shaped spring had been redesigned (turned into a more-conventional bow-shape).

The McLane hammer was sold to the Champion Blower & Forge Company in 1911, and appeared in an illustrated catalogue of January 1, 1912.[89] The machine became, naturally, the Champion

"Patented" Power Hammer. It was first offered in one size, with a 65-pound ram, for $150. Within a year or two, a 30-pound model was added, and the company emphasized its utility to carriage makers. It was noted that "the friction for stopping the hammers is such that the operator can strike one blow and stop with the hammer on the up-stroke," a nice feature which helped the hammer to further assume the role of a human striker. It was further claimed that a tire as small as three feet in diameter could be easily welded, and that tires four inches wide and one inch thick were within its capacity.[90]

The McLane hammer had passed into excellent hands indeed. The Champion Blower and Forge Company was founded in the 1870's, building hand-operated blacksmiths' blowers. The inventive skills of H. B. Keiper were complemented by the sales acumen of his brother L. B. Keiper. Their factory in Lancaster grew rapidly; and by the time the hammer was added to its line, it

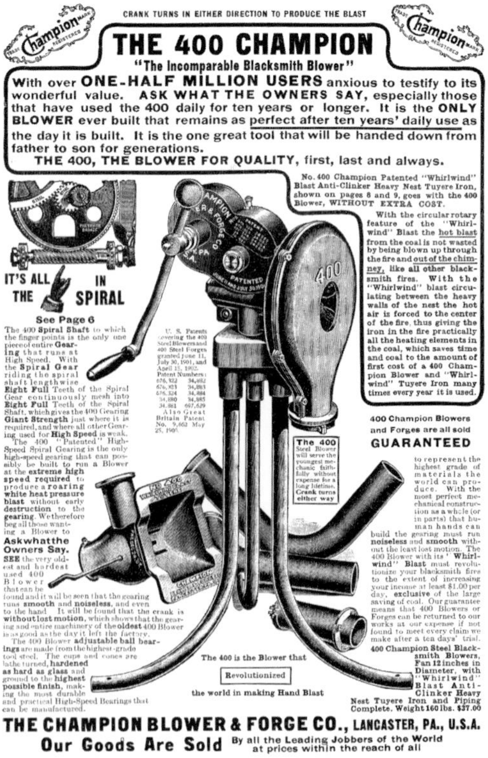

occupied more than twelve acres. The company produced a great many styles of forge blower, but by 1911, none was more successful than its No. 400. Over half a million of these tools had been sold to blacksmiths throughout the world, and it was rightfully described as "incomparable."

The hammer design remained basically unchanged through at least, 1920; though the machine had been renamed the Hercules "Patented" Power Hammer. By that time, the hammer was available with its own drive motor, an option that doubled the cost of the No. 1 (65-pound) hammer from $200 to $400.[91] The Champion Blower & Forge Company eventually redesigned its

A Hercules "Patented" Electrically Driven Power Hammer, ca. 1920, still retained the two-sided foot-treadle. By the mid-1930's, that feature had been abandoned, and a larger, 125-pound hammer had been added to the line. (*Champion Blower & Forge Company catalogues*)

hammer, eliminating the double foot treadle, in favor of a single, wrap-around design. A third, 125-pound size was also added.[92] The Champion Blower & Forge Company remained listed in the *Thomas' Register of American Manufacturers*, under "trip" hammers, through 1960.

* * *

Arthur Alfred Koch was another of the Mayer brothers' contemporaries who built a vertically-configured, guided-ram hammer. He began building hammers in Montezuma, Iowa, in about 1903. His first hammer was the distinctive Modern Power Hammer, a non-vertically-configured machine. When he left Grinnell, Iowa, in 1904, the Modern remained there, to be

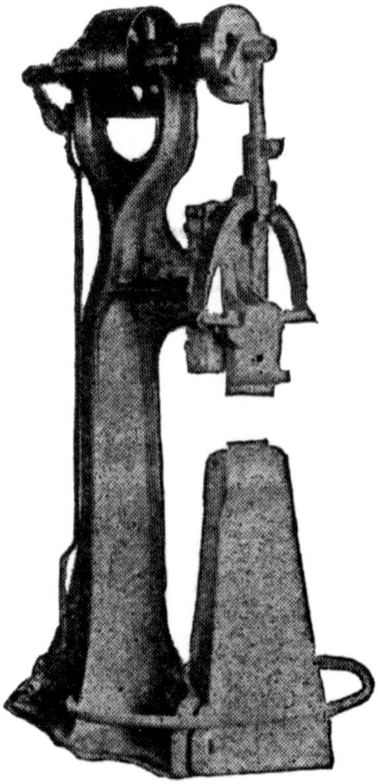

The Vulcan Power Hammer was manufactured briefly in Mason City, Iowa, by Arthur Alfred Koch. (*The Blacksmith & Wheelwright*)

manufactured by his ex-partners. Koch moved on to Mason City, where he bought the foundry and residence of Henry Hathorn; who was coincidentally embarking upon the manufacture of his own helve hammer. Koch's Vulcan Iron Works soon offered a hammer to compete with the New Little Giant. The frame of the Vulcan Power Hammer had a decidedly delicate, gothic appearance; and it likely proved less than satisfactory for heavy pounding. At a tempting price of $80, though, the hammer was probably a modest

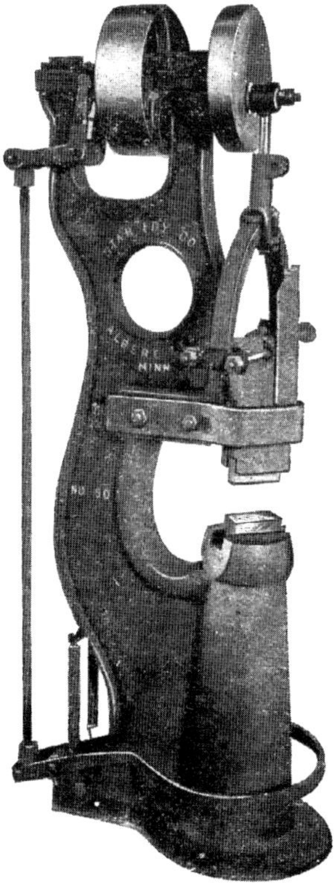

The Star Power Hammer. (***Blacksmith & Wheelwright***)

success.[93] In 1907, Koch and his partner Fred Wells moved to Albert Lea, Minnesota, establishing the Star Foundry Company. Like the Mayer Brothers Company, the Star Foundry accepted a wide range of manufacturing and repair work. It also brought out a redesigned guided-ram hammer which was known as the Star Power Hammer. Prospective customers were enjoined to "hooker to your

The Star Foundry Company building in Austin, Minnesota, ca. 1913. (*Photograph courtesy of Darlene Koch Deason*)

ingin." The company relocated briefly to Waterloo, Iowa, in 1911,[94] and Austin, Minnesota, in 1913 (where it advertised the "Starite Power Hammer").[95] Soon it returned to Albert Lea, where the hammer was then built until the mid-twenties, when the company failed.

The final chapter of the Star hammer's history was written in Texas. Koch moved to San Angelo; and in the Fall of 1926, established the San Angelo Foundry and Machine Company with new partners. Koch contributed shop machinery, working Star hammers, and patterns and jigs for the hammer line, valued at $40,000.[96] Advertisements for the relocated Star hammer boasted thousands in use, and offered four sizes.[97] While the Star hammer broke no new ground, it was well-priced. The Berger & Carter Company of San Francisco sold the 50-pound Star in 1918 for $125, while at the same time offering the 50-pound Little Giant for $180. 30-, 40- and 60-pound Star hammers were also available at that time.[98] At least one 200-pound Star hammer was evidently built, for it was included in the inventory of Koch's machinery when the San Angelo Foundry & Machine Company was formed. Within

The Star Power Hammer was built in four different cities, the last being San Angelo, Texas. (*The Blacksmith and Wheelwright*)

several years, Koch had left San Angelo, and the hammer was quickly discontinued. Koch ultimately moved to Arkansas, where he worked as a blacksmith. He died in 1956.

* * *

The MacGowan and Finigan Foundry and Machine Company of St. Louis, Missouri, came upon the scene in 1903. By July of 1904, it was advertising its Perfect Power Hammer in *The Blacksmith & Wheelwright*, featuring a special fixture for sharpening harrow and plow discs.[99] The design of this fixture was patented by Henry C. Lower, of Abilene, Kansas, who was granted patent number 833,969 on October 23, 1906 (assigned to MacGowan and Finigan). A stated objective of this fixture was to enable such discs to be sharpened without the wasting of material that would occur with grinding. The device not only helped the smith to preserve the shape of the disc, but also extended its useful life. The process of sharpening discs at the forge was a challenge, since the bowed disc had to be heated and forged in sections. For this operation, an oval-faced die was preferred and a fixture to locate the center of the disc was almost a necessity.[100] Cold, non-forging, processes ultimately won the day, with devices such as the

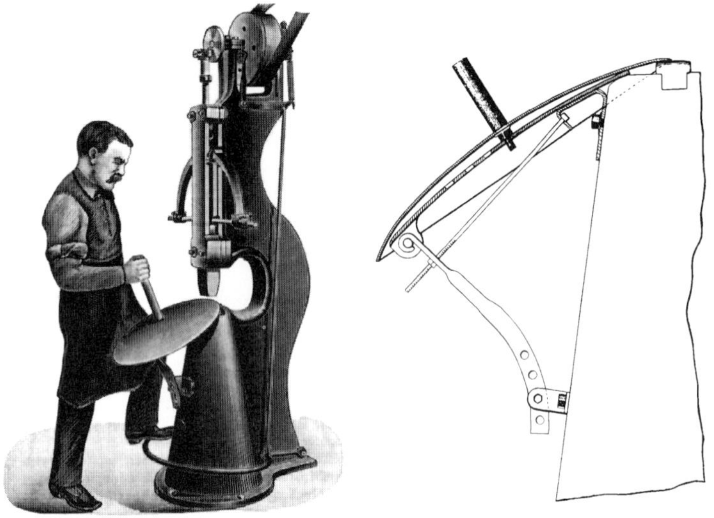

A Perfect Power Hammer in use. Note the long guides and the extremely small crank-plate (characteristic of the earliest Macgowan & Finigan hammers). The plow-disc sharpening fixture was patented in 1906, by Henry Lower. (*The Blacksmith & Wheelwright*)

Cyclone Disc Sharpener, put out by the Mayer Brothers Company, and the Star Disc Roller, manufactured by Koch's Star Foundry Company. While disc-sharpeners were sidelines for these two firms, James Depew, of Loup City, Nebraska, made a specialty of such machinery (and building hammers was his supplemental line).

The Perfect hammer itself was a lightweight version of the Belden design, which at that time was still being marketed as the Scranton hammer. Like the Scranton, the Perfect had very long guides and the lower end of the pitman (which held the bow-shaped spring) was guided like the ram. The ram was connected to the spring by adjustable steel rods, and a friction clutch within the drive pulley was said to produce from the lightest tap to the heaviest blow.[101] Within a year it was offered in three sizes: 30, 40 and 80 pounds.[102] On September 10, 1907, Peter Macgowan was granted patent number 865,787 for improvements to the Perfect hammer. Evidently, the solid one-piece links (from the ends of the bow-spring to the ram) had proven susceptible to fatigue and failure. Macgowan's solution entailed multi-part flexible links, with

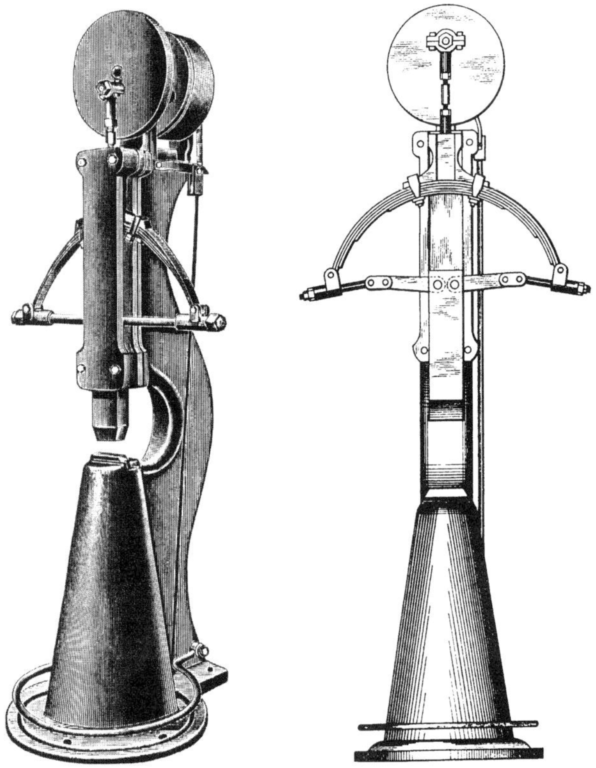

By 1906, the Perfect Power Hammer was built with a large crank-plate. (*The Blacksmith & Wheelwright*) Further improvements (notably flexible links to the ram) were patented by Peter Macgowan.

adjustment screws at each end of the bow-spring for varying the tension. Perfect hammers with this improved design appeared on the market immediately. The builders of the Perfect hammer became the MacGowan and Finigan Cordage Company in 1915, and continued to advertise under "power hammers" in the *Thomas' Register of American Manufacturers* through the mid-1920's.

* * *

Another highly successful contemporary of these machines was the Boss Power Hammer, developed by George E. Davis, of Dubuque, Iowa. As manager of the Novelty Iron Works, Davis

The Boss Power Hammer was one of the most colorfully advertised machines of its type. (*The Blacksmith & Wheelwright*) The salient features of the design were patented by George E. Davis in 1903. Details from that patent show its economical use of steel pipe and plate, as well as the simplicity of its assembly.

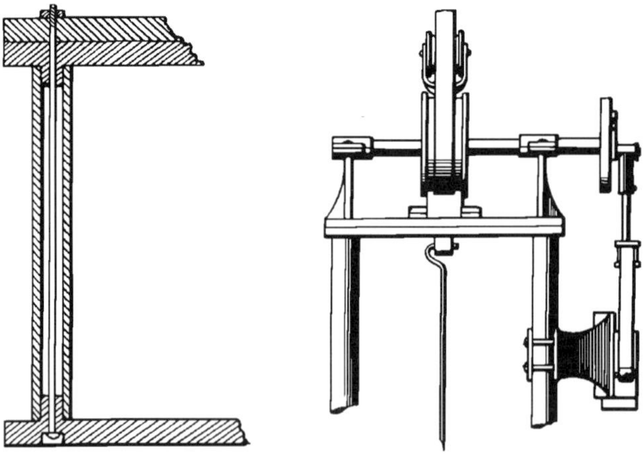

patented and built both the Boss and a small helve hammer, the Helva Hard Hitter. The design of the Boss was granted patent number 726,879 on May 5, 1903. Its novelty lay in the overall structure, which could be "readily taken apart or assembled, thus...readily transported in a 'knockdown' shape."[103] In lieu of a heavy, cast frame, the Boss used steel plates and tubes which could be securely bolted together. This indeed simplified both manufacture and shipping, and thousands were sold throughout the world. Forges in Canada, Mexico, Puerto Rico, Sweden, New Zealand and Japan were early purchasers. Davis lamented that he had not taken out a patent in England, where the Boss was quickly copied. In Japan, the hammer was said to be used for pounding fine gold and silver bars into sheet.[104]

The Boss appeared at about the turn of the century, and by February of 1902, it was being built in batches of 50.[105] "The helper with no bad habits," was a success. J.M. Weise, of Victor, Kansas, claimed to have singlehandedly sharpened nine 14" plow lays in twenty minutes.[106] This was quite a feat, for H. R. Turpin, of Missouri, proudly boasted that he could sharpen a 16" plow lay in ten minutes on his Kerrihard hammer.[107] Another Boss user claimed to have sharpened 112 lineal feet of plow lays in one day.[108] A Kansas blacksmith vouchsafed for Davis' moral claims for the

The "helper with no bad habits" did not really need the cow for mechanical stability. (*The Blacksmith & Wheelwright*)

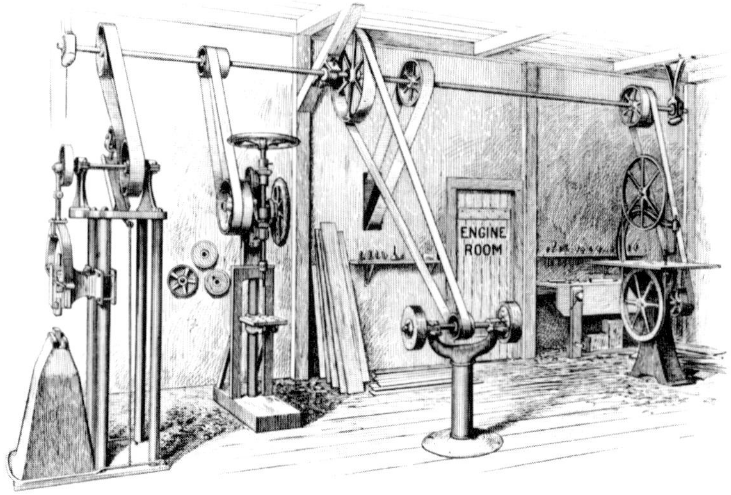

The Boss Power Hammer in an up-to-date shop. (*The Blacksmith & Wheelwright*)

Boss: "It is the best helper on earth. Never gets tired and don't run to the joints to bowl up."[109] Soon it was being built in batches of 100, and the smith was given the hint that it might sell-out come Spring. The manufacture of the Boss hammer was assumed by the Armstrong Manufacturing Company, engine-builders of Waterloo, Iowa, in 1912 or 1913.[110] By 1915, the hammer was being built by a different Novelty Iron Works in Dyersville, Iowa. The hammer was but one item in its product line, which included Victor shears, the Novirowork key-seater, the Triumph shingle machine, and the Helva Hard Hitter Helve Hammer. The New Boss Trip Hammer featured a "friction drive" arrangement which used a friction disc instead of the previously used idler pulley to control the operation of the hammer. A 75-pound hammer was added to the line, which had previously boasted only 25- and 50-pounders. The anvil of the 75-pound hammer was secured to the frame columns by a bracket, while the anvil on the smaller hammers stood free. For those smiths who had not yet added power to their shops, a special package could be ordered which included a 50-pound Boss hammer with a three horsepower gasoline engine directly connected to the base, all for $185.[111] After 1917, the Boss hammer vanished from the advertising columns of *The Blacksmith & Wheelwright*, only to reappear in a solitary advertisement, in 1923, as the product of the Eureka Manufacturing Company, of Belmond, Iowa.[112]

* * *

Mankato's Little Giant was not the only giant on the block: The Cortland Foundry & Machine Company, of Cortland, New York, built its own Cortland Little Giant Power Hammer. The machine was introduced to the readers of *The Blacksmith & Wheelwright* in 1904,[113] presumably to the consternation of the Mayer Brothers. Recommended "for carriage, forging and novelty

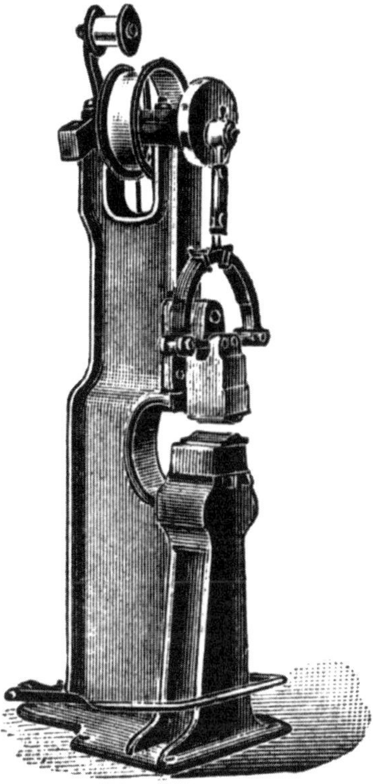

The Cortland Little Giant and the Coleman Little Sampson. (*Blacksmith & Wheelwright*)

shops," the hammer, like the Mayer brothers' Little Giant, was built in one, 50-pound, size. Since the name "Little Giant" was widely applied to blacksmithing and industrial equipment by many companies at the time, it is unlikely that Louis and Lorenz could do anything about such an infringement (Phineas Jones, of Newark, New Jersey, for example, offered the Little Giant hub-borer to wheelwrights;[114] and Sparta, Illinois, was the home of the Little Giant Punch and Shear Company[115]). The Cortland Little Giant used an uninspired but effective bow-spring linkage and was advertised in the pages of *The Blacksmith & Wheelwright* for only about a year. It was said to require about two horsepower (a lot for a machine of its size), and was probably not as successful as Cortland's other hammer, the Victor Foot Hammer, which had been patented back in 1890.

The Little Sampson Trip Hammer, which appeared in 1907, was also probably not much competition for the Little Giant. However, it boasted that it could run on just one-half horsepower, and used a Dupont-inspired, single coil-spring linkage. Built by the Coleman Iron Works Company of Elmira, New York, it claimed a price "within reach of all users."[116]

* * *

W. W. Leonard, of northeastern Minnesota, posed for a photograph of his shop, ca. 1909. To the right of a picture of President Taft stands a Minneapolis Iron Store Giant. Note the distinctive steel spring and links to the ram. (*The Blacksmith & Wheelwright*)

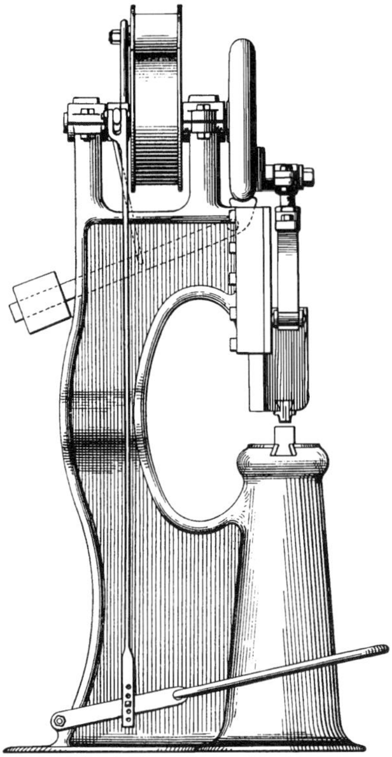

Hermann Dittbenner's patent of 1899 described a distinctive round-edged crank-plate. This functioned as a brake surface, engaged by a weighted arm when the foot-treadle was released.

The Iron Store Giant Power Hammer was another colossus which appeared briefly in the power hammer market. Built by (or for) the Minneapolis Iron Store Company of Minneapolis, Minnesota, this machine was based on the patented design of Hermann G. Dittbenner. Granted patent number 639,538 on December 19, 1899, Dittbenner's linkage featured a bow-spring which was connected to the ram by compound links. These included short, non-flexible sections and longer, thin flexible links. "These links I find to be more durable and to give greater resiliency to the hammer than links that are solid and inflexible, or leather straps," he claimed. The most noteworthy aspect of Dittbenner's design was the brake, which bore upon the distinctive, rounded edge of the crank-plate. The Minneapolis Iron Store Company soon quit the hammer-building business, but continued to sell such machines by other makers. A company catalogue from 1906 includes seven makes of power hammer (including three vertically-configured, crank-actuated hammers: the Little Giant, at $160, the Boss, for $145, and the Perfect, for $150). But by this date, its own Iron Store Giant was no longer being built.[117]

* * *

The Massillon Foundry and Machine Company of Massillon, Ohio, was best known for its steam hammers. Early in 1908, it bought-out the Lane Tool Company, of Cleveland; and it continued to manufacture and improve the steam hammers developed initially by Fredrick E. Lane. In fact, Lane himself remained involved for a time, sharing a patent in 1911 with Albert C. Creter of Massillon (assigned to the Massillon Foundry and Machine Company, it was concerned with adjustable guides for an open-frame, direct-acting steam hammer). The date of the first appearance of the crank-actuated Massillon Belt Power Hammer is not known, but the company initiated a listing under "Hammers, Power" in the *Thomas' Register of American Manufacturers* in 1916 (to accompany its established listing under "Hammers, Steam.").[118]

The Massillon Belt Power Hammer was manufactured in three sizes, with rams weighing 50, 77 and 104 pounds. The design was most conventional: a Shaw style bow-spring linkage with plain, non-adjustable toggle arms. An idler pulley was used to tension the drive belt and thereby control the hammer. A T-slot on the crank-plate permitted some adjustment for various sizes of work. The anvil was integral to the frame, and gracefully styled.

The Massillon Foundry and Machine Company was located in the notorious Columbia Heights area, just outside of Massillon. A short walk from the foundry, one could enter any number of so-called coffee houses "filled with the stench of tobacco smoke, cheap perfume, garlic, tough men, cheap women and rotten booze."[119] Alongside heavy industry in Columbia Heights, the ethnic

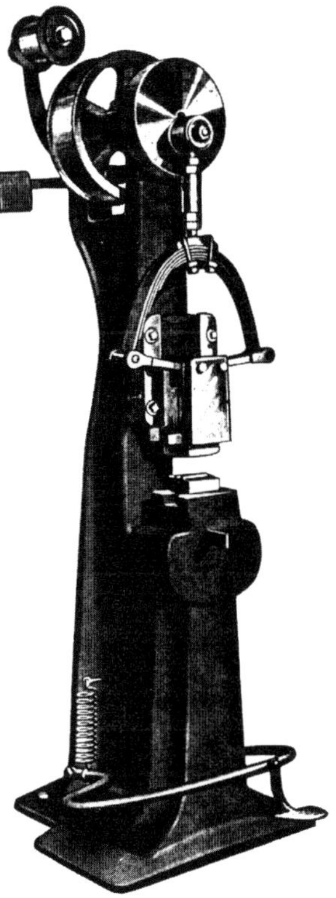

The Massillon Belt Hammer. (*Machinery*)

melting-pot of America in the 1920's was at full boil! Massillon City Directories from this time include the Massillon Foundry and Machine Company, listing I. M. Taggart as president, J. C. Haring as vice president, W. J. Cadman as secretary and assistant treasurer, and F. G, Harrison as treasurer and general manager. In addition to hammers, the company offered punches, shears, presses and machine tools.[120] The company vanished in the late 1920's, but the hammers may have been produced briefly thereafter by the Massillon Power Shovel Company (a solitary listing for this firm appears in the 1928-29 edition of the *Thomas' Register of American Manufacturers*).[121]

*   *   *

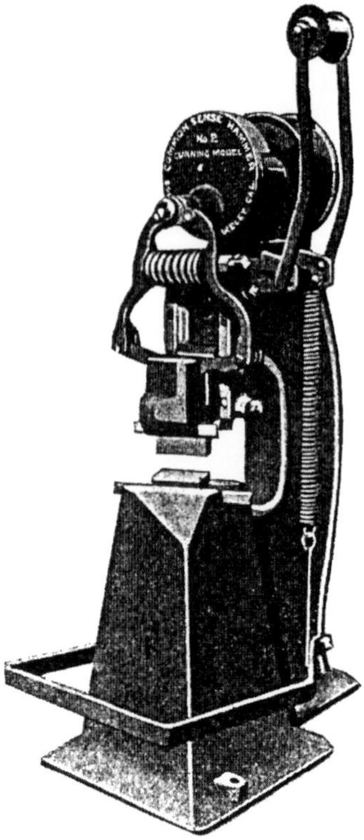

The Schuyler Common Sense Power Hammer. The meaning of the designation "gunning model" on the crank-plate is not known. (*Waterhouse & Lester Company catalogue*)

While the vast majority of power forging hammers were manufactured in the Midwest and the East, several were built in the West. The Schuyler Company of Berkeley, California, offered regional competition to the Little Giant hammer with its own Common Sense Power Hammer. This machine employed a Dupont-style linkage, with short, stout bars connecting the toggle-arms to the ram. An idler pulley was used to control the hammer's operation in all three sizes (50, 75 and 100 pounds). Waterhouse and Lester, a machinery dealer in San Francisco, offered both the Common Sense hammer and the Little Giant in its 1924 catalogue: a 50-pound Common Sense ran $275, while the comparable Little Giant cost $365.[122] It is thus not surprising that the Common Sense hammer was popular on the West coast, and a significant number of these hammers are still in use there today.

Details of the Schuyler Company are sketchy. The company began in San Francisco in 1905 as a machine shop. The principals were John B. and Frank B. Schuyler. Unfortunately, this was not an auspicious time to locate in San Francisco, and the enterprise was soon closed by the great earthquake of 1906. In 1907, the company reappeared, across San Francisco Bay, in Berkeley. The Schuyler Company was listed under "power hammers" in the *Thomas' Register of American Manufacturers* from 1914 through 1925. Little is known of the company's first decade, but by the end of World War II, Frank B. Schuyler was specializing in the production of his Common Sense Washing Machine. In 1922, Schuyler purchased the Johnson Electric Washing Machine Company of Oakland, California. The Berkeley plant of the Schuyler Company was then closed and moved to neighboring Oakland.[123] Production of the Common Sense Power Hammer apparently continued there for several years.

Inland, Colorado blacksmiths could support their own state's industry by purchasing a Denver-built Flint-Lomax Power Hammer. This hammer employed a bow-shaped spring linkage with connecting links to the ram. The stroke was adjustable through a slot in the crank-plate, and a variable pitman facilitated adjustment for different thickness of work. The Flint-Lomax Power Hammer had a graceful frame casting which was distinctively hollow behind the dies for the extension of long workpieces. The builder, the Flint-Lomax Electric & Manufacturing Company was incorporated in 1900, and was listed in the *Thomas' Register of American Manufacturers* under "power hammers" from 1907 to 1920. In 1913, the corporation was changed to simply the Flint Electric & Manufacturing Company; and, in 1937, the corporation was dissolved. The full extent of its hammer line is unknown, but a 1917 catalogue of the Hendrie & Bolthoff Manufacturing and Supply Company, of Denver (known affectionately to its customers as "Hoptoads and Bullfrogs"), included a 65-pound Flint-Lomax hammer (described as its "Standard Power Forging Hammer").[124]

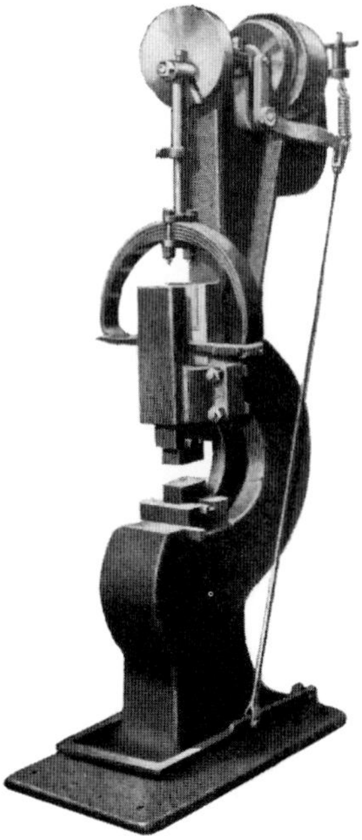

The Flint-Lomax Power Hammer was built in Denver. (***Hendrie & Boltoff Manufacturing & Supply Company catalogue***)

* * *

As industry flourished on the West coast, so did innovation. The Radial Power Hammer Company of Los Angeles, California, was formed to develop a novel, open-die power forging hammer which created a versatile work-station for the blacksmith. The Universal Radial Power Hammer was unique in offering forging, upsetting and swaging stations to which the hammer mechanism could be rotated. In its central position, the hammer was used for normal, flat-die work, to the user's left the hammer became a dedicated upsetting machine, much like the Little Giant Neckyoke Singletree Hammer. On the right-hand side was a large, circular swage block which could be rotated to a wide variety of profiles. This conveniently replaced the jumble of individual hardies that would be positioned in the square "hardy hole" of the blacksmith's regular anvil, the use of which required a human striker. The

The Universal Radial Power Hammer in operation. The smith has rotated the hammer mechanism to the right-hand station, where he is using the swage block. (*Joseph T. Ryerson Co. trade literature*)

central hammer mechanism was rotated easily via a bevel gear, and was quickly locked into position. The hammer mechanism itself was unusual in that the bow-shaped spring was attached directly to the crank-pin, without an adjustable pitman. That limitation was at least partially overcome by making the entire hammer mechanism vertically-adjustable.

The Universal Radial Power Hammer is first described in patent number 953,768 granted April 5, 1910, to Fred E. Sutherland, of Los Angeles. The patent was assigned to the Radial Power Hammer Company, which had incorporated on September 7, 1909. A second patent, number 973,620 was granted to Sutherland and James A. Brown on October 25, 1910. The president of the corporation was a well-known consulting engineer, Henry E. Brett. Brett had been superintendent of construction of Los Angeles'

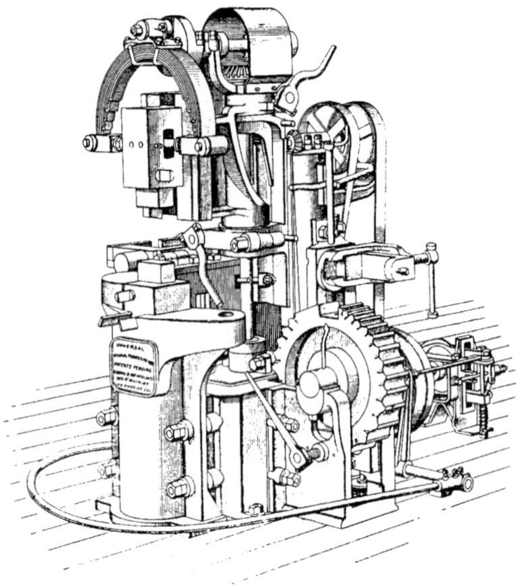

An illustration from the Sutherland and Brown patent of 1910 shows the rotating swage block. Note the direct connection of the bow-spring to the crank-pin.

electric railway in 1890. He also designed various types of machinery, including the device that stamped the Diamond Walnut logo on each of its nuts (no mean feat!).[125] The hammer was unveiled in the March, 1910, issue of *Machinery* magazine, and offered in one size having a 75-pound ram.[126] The Radial hammer was subsequently offered by the Joseph T. Ryerson Company, of Chicago, Illinois, as the Ryerson Radial Power Hammer, though Ryerson never participated in the manufacture of the machine.[127] As part of its extensive line of forging hammers (which included the Little Giant, the Modern, the Rochester helve, and the Bell steam hammer), the Radial was offered in four sizes, including 50-, 75-, 125- and 200-pound models.[128] Despite its versatility, the Radial Power Hammer was soon discontinued. Prices for the Radial hammer are not known, but they were likely substantial. The machine was a heavy, complicated industrial hammer, and was probably well beyond the means of most independent smiths. The corporate status of the Radial Power Hammer Company was suspended by the Franchise Tax Board on November 30, 1913.

*   *   *

The Radial Power Hammer was not the only machine to combine several hammers in one: The J. D. Foot and Power Hammer Combined was one of the more remarkable hammers to be built during the early decades of the twentieth century. Produced for the J. D. Hammer Company of Sabetha, Kansas, the J. D. hammer consisted of a crank-actuated guided-ram hammer, which could be run off of a smith's line shaft. Additionally, though, the bow-spring linkage could be disengaged and replaced by a direct-actuated link to the foot-treadle, so that the hammer could be foot-powered for single blows! This was a splendid idea; and the

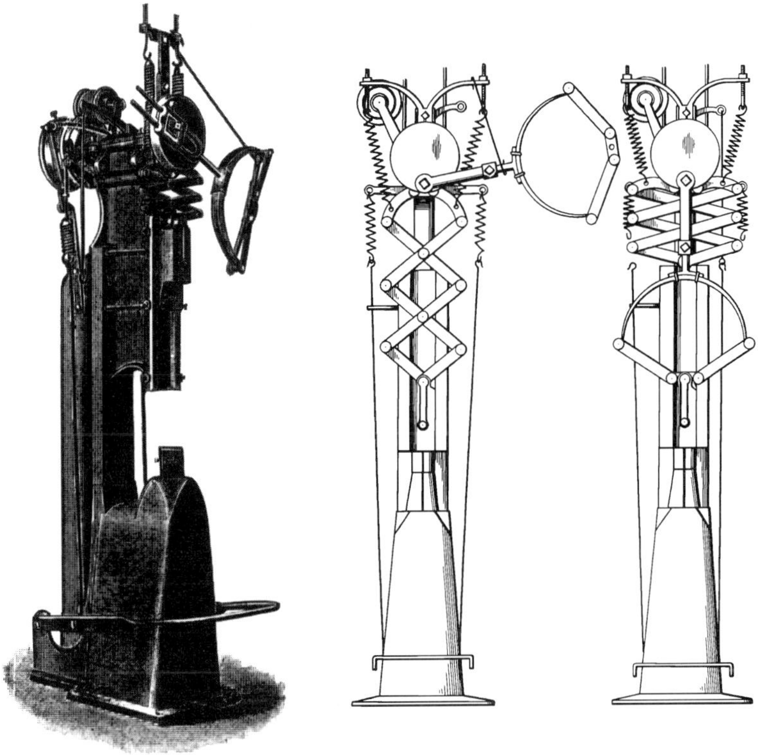

The J. D. Foot and Power Hammer Combined was based upon the 1903 patent of Benjamin Jinks and Edwin Davis. With the bow-spring linkage connected, the hammer was a conventional belt-driven power hammer. With the linkage disconnected, the hammer could function as a single-blow Oliver! (*The Blacksmith & Wheelwright*)

inventors, Benjamin F. Jinks and Edwin W. Davis, were granted patent number 736,265 for their innovation on August 11, 1903. A number of subscribers to *The Blacksmith & Wheelwright* reported using this ingenious machine. Clark Olds & Company of Nebraska, had used its J. D. hammer for a year and a half. The foot-power capability pursuaded them to buy it; and while they observed that it cost more than many other makes, they regarded it as "well worth the difference in price."[129] A J. D. hammer stood proudly in the center of the shop of C. H. Wood, of Wisconsin. He reported that he had been in the trade for 43 years, and was always willing to learn something new.[130] The circumstances surrounding the manufacture of this hammer are uncertain: the crank-plate of the J. D. hammer proclaimed that the machine was manufactured *for* the J. D. Hammer Company. The identity of the actual builder is not known. Advertisements for the J. D. hammer first appeared in *The Blacksmith & Wheelwright* in the summer of 1906, but soon disappeared.[131]

* * *

Despite the advantages and popularity of vertically-configured guided-ram hammers, American inventors also explored designs which placed the crank behind the work area. Many different machines of this type were manufactured, and some were among the most innovative of all power forging hammers. The following chapter describes these non-vertically-configured guided-ram hammers, and their builders.

The helper with... bad habits?

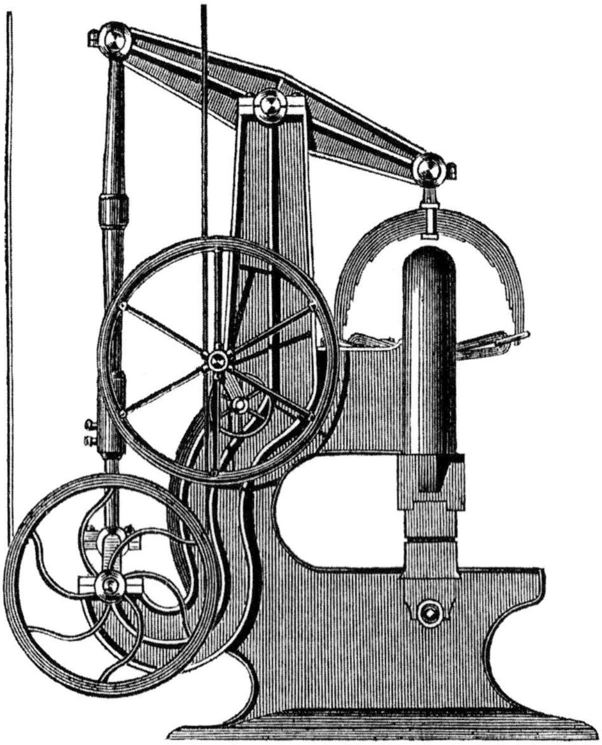

The 1,000-pound model of the Shaw and Justice Dead Stroke Power Hammer was built in a non-vertical configuration. (*The Scientific American*)

CHAPTER 5

"The Hammer that Pounds Out the Profits"

Many crank-actuated, guided-ram hammers were built in a non-vertical configuration. These machines almost always employed a pivoting beam or helve-like element to communicate a reciprocating motion to the ram. The crank was then located at the back of the machine. While this arrangement generally required more shop floor-space than was needed for a vertical configuration, there were advantages, and a number of factors accounted for their on-going development. In some cases, height was an over-riding consideration. In other cases, a non-vertical configuration provided greater options for adjustment of the hammer's stroke. While many non-vertically-configured hammers used spring linkages similar to those found on vertical hammers, others used entirely different means to deliver powerful blows without jarring the machine. Placement of the crank behind the work area gave non-vertically configured hammers some of the advantages of helve hammers; but the guided-ram eliminated the main problems associated with the helve.

In a sense, the first crank-actuated open-die power forging hammer, Hotchkiss' Atmospheric Hammer, was non-vertically configured. The placement of the crank behind and below the air-cylinder spring linkage was done for some of the same reasons that dictated a non-vertical configuration in later machines; however, the beamless mechanism of Hotchkiss' hammer was most unusual. In fact, subsequent atmospheric hammer designs were either vertically configured (the Glossop / Hackney hammer, for example) or employed a pivoting beam (as in the patented designs of John Butterfield). Since Hotchkiss' hammer has already been discussed at length in Chapter 2, it will not be dealt with further here. One of the only additional non-vertically-configured hammers which did not employ a pivoting beam was the so-called Pillar Hammer built by Beaudry, of Boston, Massachusetts. Though such a beamless design was unusual, it nonetheless fits comfortably into this category. Perhaps future designs in non-vertically configured guided-ram hammers might also explore ideas which omit the pivoting beam.

Height and stability were undoubtedly considerations in the non-vertical configuration of certain hammers. This was true of Hotchkiss' hammer, as well as the largest sizes of the Shaw and Justice Dead Stroke Power Hammer. What was practical for a hammer with a lightweight ram became less so for a hammer with a 1,000-pound ram. The largest, 500- and 1,000-pound Shaw and Justice machines would have been prohibitively tall in a vertical configuration, and an increase in the mass of the base would have been required for stability. Such concerns would also have been magnified as increasingly-large crank-actuated hammers of this type were contemplated (remember that the manufacturer claimed to be planning to build hammers as large as ten tons). The pivoting beam on the largest sizes of the Shaw and Justice hammer was also mechanically advantageous, for it reduced the power required to operate the machine. It should be noted that it was possible to build a very large vertically-configured crank-actuated hammer, as proven by the 1,000-pound model "V" Little Giant. However, the practicality of this machine is open to debate; and, in fact, it was built in very small numbers.

The compact nature of vertically-configured hammers limited the extent to which the stroke could be modified. It was generally possible to adjust such hammers for varying thicknesses of work by altering the length of the pitman which connected the crank to the spring. The strength of the blows could be altered by changing the position of the pin on the crankplate (as with Belden's improved version of the Shaw and Justice hammer) or by adjusting the tension on the toggle arms (as with the Little Giant). Neither of these adjustments could be accomplished while the hammer was in operation, nor could they be done easily enough to respond to continuously changing work conditions. Instead, the force of blows with many vertically-configured hammers was increased or decreased as a function of the rate of blows per minute. The faster the crank rotated (usually as a tightened idler pulley reduced drive-belt slippage), the harder the hammer's blows. Unfortunately, this was not necessarily the most desirable combination! If a blacksmith required rapid, light blows on a thin workpiece, or heavy, deliberate blows on a large forging, a vertically-configured hammer might not be suitable. Some non-vertically-configured hammers were similarly limited, but others made use of the additional pivoting beam to facilitate adjustment of their stroke without necessarily changing the rate of the blows. This adjustment was usually made with the hammer idle, by changing the point at which the pitman was connected to the beam. Several hammers used a non-vertical arrangement to make it possible to adjust the stroke even while the machine was in full operation. The Modern Power Hammer, designed by Arthur Alfred Koch, employed a movable fulcrum to enable adjustment of the force of the blows without changing the number of blows delivered per minute. Other hammers featured the capability of raising or lowering the fulcrum, while the machine was running, to accomodate work of varying thicknesses.

A non-vertical configuration permitted some interesting alternatives to the spring linkages of Shaw and Dupont. Joseph Palmer's Power Spring Hammer used flat steel leaves to make the pivoting beam itself the flexible element. Early, trip-actuated hammers had relied upon the flexibility of their wood helves to absorb some of the shock of impact. The Palmer hammer made a success of this approach. Increasingly elaborate spring-cushioned beams appeared in non-vertically configured hammers. This idea culminated in the sophisticated design of Henry Alinder which appeared in the second decade of the twentieth century.

By placing the actuating crank of a power forging hammer behind the work area, greater clearance existed above the hammer ram. For certain operations, such as the welding of wagon and buggy tires, this was particularly useful. Helve hammers, of course, provided the maximum clearance, with no obstructing mechanism above the ram. Despite its Dupont-style spring linkage, the tiny Kerrihard Hammer was hailed as being especially suited to tire welding. Even a relatively small tire could encircle its head so that the tire's ends could be forge-welded together. The Kerrihard Hammer (which sold for $60 in 1909) illustrates another factor that influenced the development of some non-vertically configured hammers: they could be cheaply built and sold at a highly competitive price.

Non-vertically configured, crank-actuated hammers can be divided conveniently into two categories: 1) those machines which have their fulcrum located near the center of the beam, between the ram and the connection to the crank; and 2) those which have their fulcrum at the end of the beam. The Palmer Power Spring Hammer is an early example of the former, and the Kerrihard hammer is an example of the latter. The majority of non-vertically configured hammers were of the first category, but the distinction is subtle. It was possible to include adjustment of the stroke in either arrangement by changing the length of the pitman or by moving its point-of-connection along the beam. Also, hammers of both types were built which had the capability of raising or lowering the fulcrum (to accommodate work of varying thickness). The centrally-located fulcrum did, however, offer several advantages. As previously mentioned, it was possible to design a hammer in which the fulcrum could be moved to adjust the stroke. More importantly, the mechanical advantage of a centrally-located fulcrum reduced power consumption and the expense of operating the hammer. For this reason, hammers with the fulcrum at the end of the beam tended to be smaller machines, while hammers with a centrally-located fulcrum were built in sizes up to 1,000-pounds.

* * *

S. C. Forsaith and Company of Manchester, New Hampshire, claimed that its Palmer Power Spring Hammer was "fully equal the best steam hammers in existence."[1] This machine

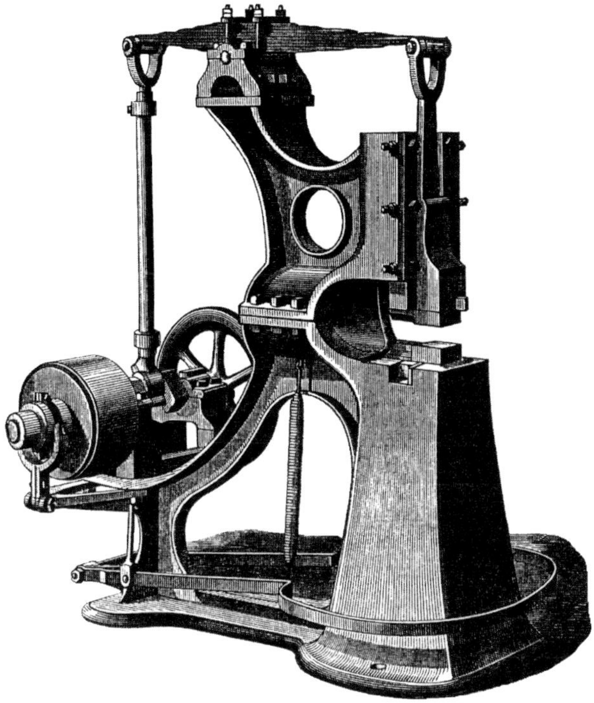

The Palmer Power Spring Hammer used a flexible beam in lieu of a spring linkage. (*The Scientific American*)

used a pivoting beam made-up of flat spring leaves, to transmit a reciprocating motion to the ram and also to provide a flexible linkage. Like Shaw's semi-elliptical spring, the beam stored energy and increased the force of the blow. Built in ten sizes, from 20 to 225 pounds (and priced from $225 to $1,050), it was relatively safe to compare the machine to larger, more costly steam hammers. Advertising literature for the Palmer Power Spring Hammer thus avoided the more appropriate, but difficult, comparison with the Shaw and Justice Dead-Stroke Power Hammer.

The Palmer hammer was designed by Joseph Palmer of Concord, New Hampshire. Palmer was a professional spring maker, and originally developed the hammer for his own use at Palmer, Ward and Company.[2] On January 9, 1872 he was granted patent number 122,647 and the machine was quickly placed into

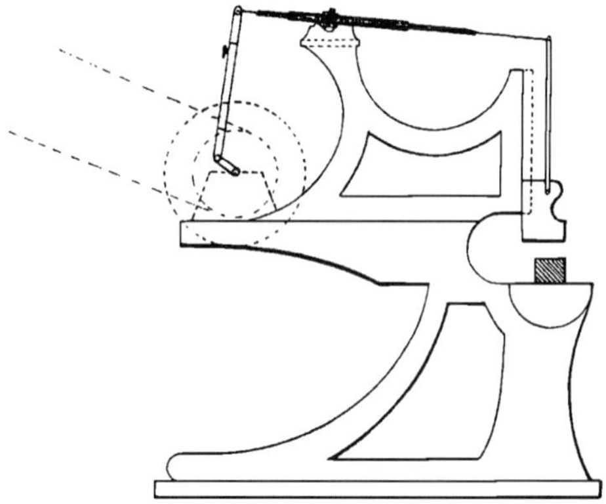

Palmer's hammer was patented in 1872 by Joseph Palmer, a practical spring maker.

production at the works of S. C. Forsaith. A profile by *The Scientific American*, in October of 1873, claimed that there were then some twenty of these hammers in operation in various parts of the country.[3] In the following month, the Palmer Power Spring Hammer was awarded a silver medal at the American Institute Fair held in New York City.[4] Testimonials indicate that the hammer was used in a variety of forging establishments. The Eagle Vise and Anvil Works of Fisher & Norris in Trenton, New Jersey, bought a Palmer hammer. They praised its friction clutch arrangement and particularly appreciated the counter-balancing of the fly-wheel, which protected the line-shaft drive from strain at the point when the blow was struck. The Palmer hammer was also popular with file and cutlery works, and was most heartily recommended by Larkin and Varnum, manufacturers of "The Kitchen Gem" vegetable fork.[5] Like the Shaw and Justice hammer, the Palmer hammer was especially effective in situations where light, rapid blows were required. A 20-pound Palmer hammer was designed to run at an impressive rate of 500 blows per minute (decreasing to 100 blows per minute for the 225-pound machine).[6] Palmer, himself, remained in the spring-making business. In the *Concord Directory* for 1880-1881, he advertised "superior tempered" springs intended for "omnibus, coach, wagon and buggy."[7] The Palmer hammer was introduced into England in the early 1880's as

Dunham's Spring Beam Hammer. It was built there in a full range of sizes by G. Booth and Company of Halifax.[8]

S. C. Forsaith and Company was founded in 1860, and rose to occupy extensive works. From seventy-five to one hundred hands were employed there in the mid 1880's.[9] Samuel Caldwell Forsaith had served an apprenticeship in the famous machine shop at the Amoskeag Mill in Manchester, Vermont, and worked in a number of other machine shops before establishing his own business. Among his early inventions and products was a highly successful machine for folding newspapers.[10] His enterprise grew swiftly. In addition to the Palmer hammer, the company built circular saw mills, woodworking machinery, line-shaft components, and hand-operated fire engines. Portable engines and circular saws were particular specialties; used steam engines and machinery were sold as well. An advertisement in *The Blacksmith and Wheelwright* from June of 1883 proclaimed that the company had 2,662 machines, new and used, currently in stock![11] In 1884 the business changed hands, incorporated, and became the S. C. Forsaith Machine Company. An advertisement in the 1897 *Manchester City Directory* mentions the Palmer Power Spring Hammer;[12] however, the company was never listed under "power hammers" in the *Thomas' Register of American Manufacturers* (which first appeared in 1905). By 1925 the company was judged to be inactive and its charter was repealed by the state of New Hampshire.[13] The Palmer hammer vanished along with the S. C. Forsaith Machine Company.

* * *

Prior to the development of its vertically-configured hammers in the 1890's, the Beaudry Company of Boston, Massachusetts, built a successful line of non-vertically-configured hammers. Alexander Beaudry was a blacksmith in East Boston, and is reported to have developed a power hammer in the early 1870's, presumably for his own use.[14] In the early 1880's, as a partner in the firm of Beaudry and Cunningham, his ideas about hammers were refined and finally placed into production. Advertisements for

Beaudry's Upright Cushioned Hammer appeared in *American Machinist* in the summer of 1885,[15] and seven Beaudry hammers were exhibited at the American Institute Fair in New York City the next year.[16] In addition to hammers, dies and forges, the company built special machinery to order, and handled the production of various products for other manufacturers.[17]

Alexander Beaudry's first power hammer patent, number 244,370 which was granted on July 19, 1881, combined a flexible, spring steel beam with a ram which was cushioned by a linear arrangement of two coil springs. The lower coil spring contributed to the downward throw of the ram, while the upper spring helped to raise the ram quickly after the blow was struck. A second patent, number 264,816 granted September 19, 1882, described a hammer with a more rigid, wood beam. To compensate for this rigidity, another similar set of coil springs was used as an element of the pitman to impart elasticity and to protect the drive mechanism. While the linear coil spring linkage may have been practical, it was apparently not used in production. Instead, the early reputation of Beaudry's hammers was built on a design that was altogether

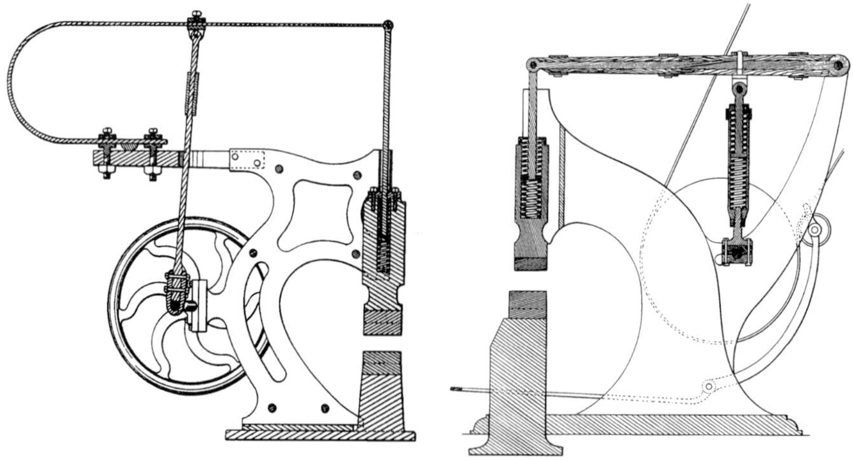

Alexander Beaudry's first power hammer patents date from the early 1880's. His design of 1881 (left) used a flexible beam, much like Joseph Palmer's earlier hammer. Beaudry's design of 1882 (right) used a fairly rigid wood beam with in-line coil-spring linkages. This was among the earliest instances of an end-pivoting beam.

different. The Beaudry Upright Cushioned Hammer used a conventional leather strap to suspend the ram from the helve. The pitman was connected to the helve by means of a rod that extended behind the helve. The rod was bolted to the helve with rubber cushions in between, thus providing the necessary elasticity and isolation for the drive.

Beaudry's Upright Cushioned Hammer. (from "Notes on Steam Hammers," by C. Chomienne in *The Railroad and Engineering Journal*)

The design of Beaudry's Upright Cushioned Hammer is described in his third patent, number 281,001 which was granted on July 10, 1883. The drawings which accompany this patent are less fanciful than those which illustrate the earlier two patents, and it is likely that production of this hammer was already underway.[18] The patented design included a buffer spring placed above the hammer ram, intended to enhance the elasticity and force of the blow. However, by the time the hammer was first illustrated in *American Machinist*, that detail had been eliminated. The actual Beaudry hammer also included a straightforward provision for adjustment of

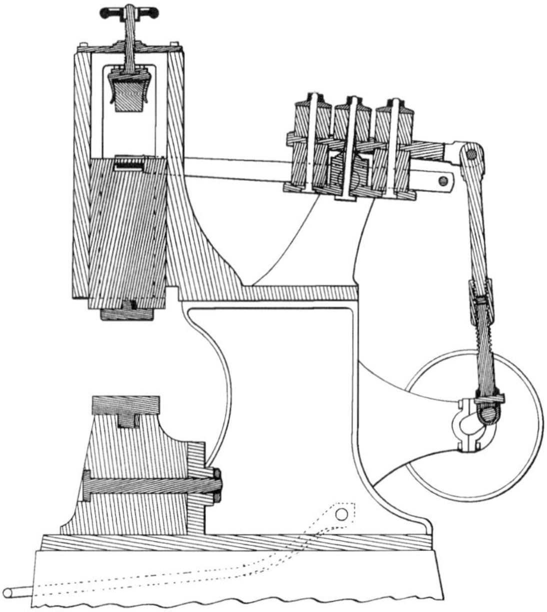

The distinctive design of Beaudry's Upright Cushioned Hammer was described in Alexander Beaudry's patent of 1883.

the stroke, with the addition of a sliding friction sleeve connecting the pitman to the cushioned extension of the helve. In what was to become a distinctive feature of Beaudry hammers, the ram was secured by guides which were oriented in-line with the helve (perpendicular to the "front" of the hammer). The hammers built by Beaudry and Cunningham were immediately successful, and spurred the Bradley Company of Syracuse, New York, to quickly develop their own non-vertically configured, guided-ram forging hammers.

In mid-1887, Beaudry and Cunningham became simply Beaudry and Company.[19] The Beaudry Duplex Press was unveiled later that year, and it proceeded to garner awards at the American Institute Fair and the Fair of the Massachusett's Charitable Mechanics Association in Boston.[20] Early in 1889, with the success of the press and the Upright Cushioned Hammers, the company moved into a larger shop.[21] A combined forging, punching, and

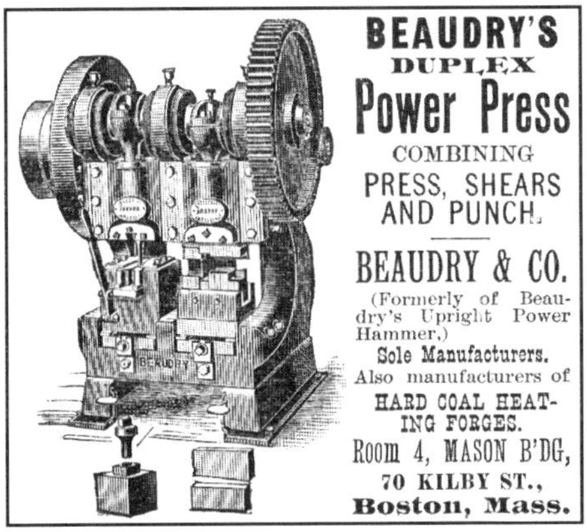

Beaudry's Duplex Power Press. (*American Machinist*)

shearing press was introduced, and was featured on the coveted front page of *American Machinist* in April of 1889.[22] In 1890, a new hammer was placed on the market. The so-called Beaudry Pillar Hammer used an arch, or double frame, configuration borrowed from much larger steam hammers. With this design, Beaudry sought to produce "a cheap and efficient hammer that will occupy a limited space and be easy to regulate and keep in order."[23] This unusual machine is described in patent number 461,917 which was granted

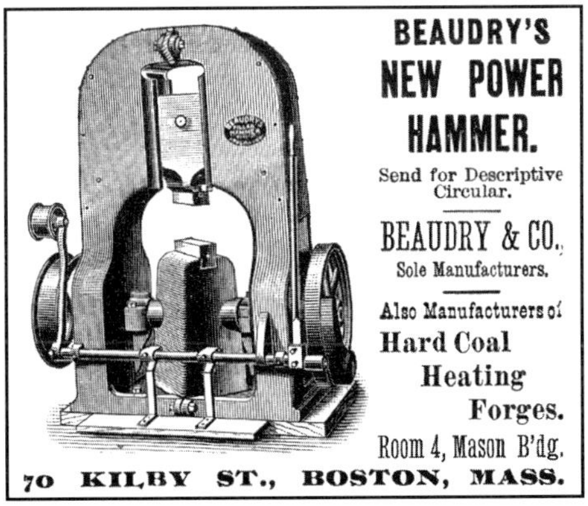

The Beaudry Pillar Hammer is one of the few non-vertically-configured crank-actuated hammers which did not employ a pivoting beam. (*American Machinist*)

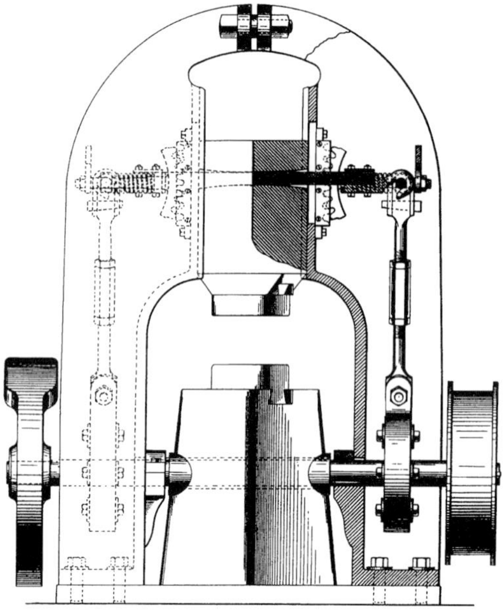

Beaudry's 1891 patent describes the design of the Pillar Hammer.

to Alexander Beaudry on October 27, 1891. Included in this patent are numerous variations on the suspension of the ram. The hammer is unique in using a double crank (one on each side of the anvil) which ran off of a single crankshaft extending through a hole in the anvil. The ram was suspended from a flexible element which was attached to a pitman on either side. The length of the pitman could be adjusted to accomodate work of varying thickness, and adjustment for wear of the guides was accomplished by simply replacing the shims used in bolting together the left and right hand castings. This was a visually elegant machine, and quite modern in its attempt to conceal the mechanism from view. It was advertised for only a brief time, and was quickly superceded by another Beaudry innovation.[24]

The Beaudry Cylinder Power Hammer was introduced in January of 1892.[25] While closer in mechanical design to the Upright Cushioned Hammer, this new machine also used a cast iron frame to enclose its mechanism. The hammer is described in

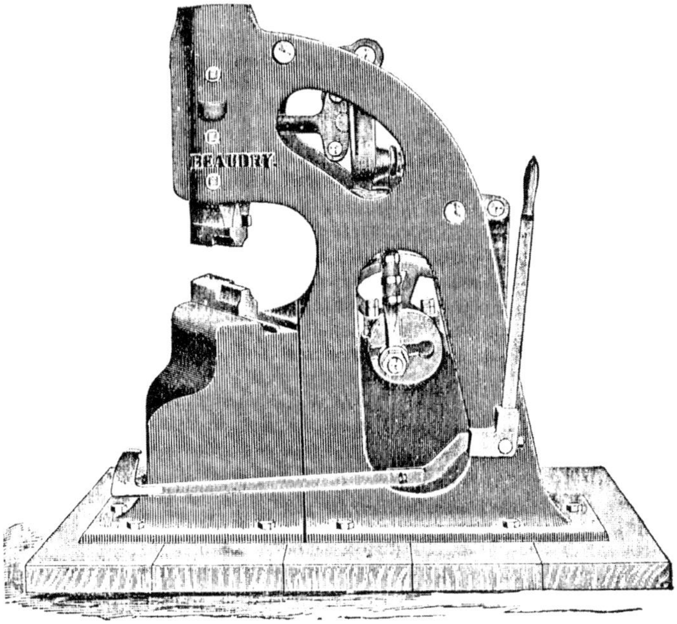

Beaudry's Cylinder Hammer took its name from the cylindrical shape of the ram. The mechanism is described in Beaudry's patent of 1892.

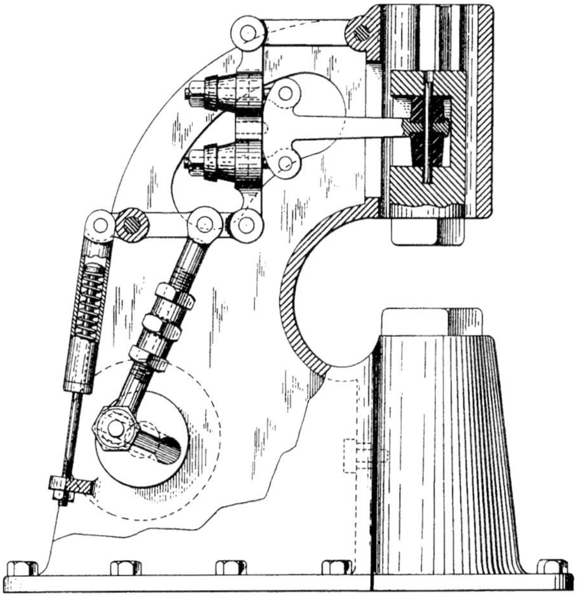

patent number 487,692 which was granted to Alexander Beaudry on December 5, 1892, and took its name from the cylindrical shape of the ram. A short helve was attached to the middle of the ram with rubber cushions above and below. The helve, in turn, was attached through rubber cushions to an oscillating element which was driven by the crank. An opening in the side of the frame permitted access to the crank; which, through a T-slot, allowed adjustment of the stroke. This hammer design was also short-lived, for the company soon embarked upon manufacture of the vertically-configured hammers that it was to produce for the next three decades. As previously described, the patents that pertain to the vertical hammers were granted to Augustin Beaudry in 1894 and 1895. Alexander Beaudry died in 1909, at the age of seventy. In 1924, the company moved its factory to nearby Everett, Massachusetts, where there was a brief flurry of activity.[26] The introduction of an air-cushioned helve hammer in 1924[27] and a direct-acting air hammer the next year suggest a company adrift and desperate to re-establish itself in the hammer market.[28] But even before the crash of the Stock Market in 1929 and the Great Depression, the Beaudry Company had passed out of existence.

* * *

The Bradley Company of Syracuse, New York, was already well-established in the power hammer market when it introduced its own two versions of a non-vertically-configured hammer. These new machines, the Upright Helve and the Upright Strap Hammer, used the same type of rubber cushions that were incorporated in Bradley's popular Rubber Cushioned Helve, but employed a very different means of communicating a reciprocating motion to the helve. In fact, these machines were quite similar to the Beaudry Upright Cushioned Hammer which had appeared on the market a few years earlier. Both manufacturers used an "oscillator," but while Beaudry bolted the oscillator directly to the helve (as an extension of the helve, with the rubber springs in between), Bradley employed a pivoting yoke. This yoke communicated its motion (through rubber cushions) to a cast iron pivot, or "husk," which gripped the helve. During the late 1880's, the exact relationship between these two companies and their upright hammers is uncertain. While they were initially fierce competitors, a Bradley advertisement in *American Machinist* from 1889 includes the Beaudry Upright![29] If there was a marketing arrangement between the two firms, it was probably short-lived.

This advertisement from 1889 indicates that there was some sort of marketing arrangement between the Bradley and Beaudry companies. (*American Machinist*)

Bradley's Upright Helve and Upright Strap Hammers were fundamentally quite similar (despite the use of the term "helve," these were guided-ram hammers, and the helve was simply a pivoting beam). They differed in the design of the so-called helve and the means used to attach the helve to the ram. The Upright Helve used a beam of hard maple which was attached to the ram with a strip of leather belting. A single rubber cushion between the helve and the ram provided snap and elasticity to the hammer blow. The result was an extremely robust hammer. The Upright Strap Hammer, on the other hand, used a leather strap to suspend the ram. This was a proven design and could be found on the hammers built earlier by Justice, Belden and others. The helve used in the Upright Strap Hammer was initially composed of steel strips with wood in between, but eventually the wood was eliminated. This design provided rapid, snappy hammer blows. With the ram moving vertically in guides, these two Bradley hammers were intended to offer the best of both worlds: a smooth, helve hammer action with precisely-aligned, parallel die faces. While the "helve" became, in a sense, an artifact, these hammers were effective forging machines and were commercially successful for approximately half a century.

The development of Bradley's guided-ram hammers was largely the work of Syracuse inventor Michael Deering. Deering's first Bradley-assigned patent, number 208,719 which was granted on October 8, 1878, proposed an alternative to the oscillating carriage of the Rubber Cushion Helve Hammer. Conceptually a step backward, Deering's design used a simple pivoting helve, but introduced a rubber-cushioned linkage to the pitman (which was attached

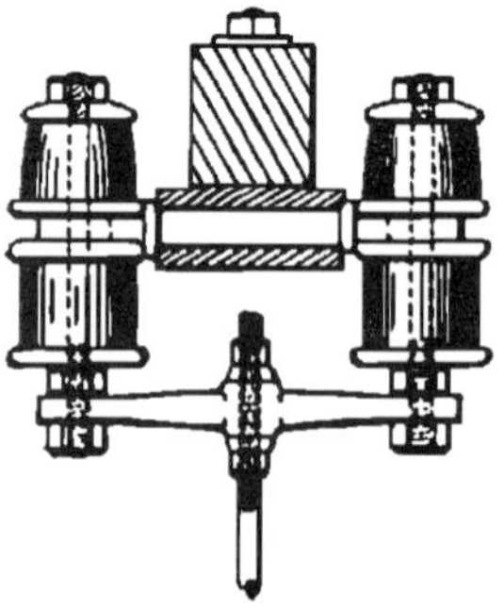

Michael Deering's patent of 1878 describes a rubber-cushioned linkage for a crank-actuated helve (the center rectangle at the top is the cross-section of the beam).

between the pivot and the head). This design was not incorporated into the Bradley line, and the patent was probably intended primarily to protect a viable alternative from the competition. But the investigation of alternatives to the oscillating carriage was useful, and with a little competitive pressure from Alexander Beaudry, it was soon nurtured to bear fruit.

The Bradley Upright Helve Hammer had a guided-ram, and was thus not truly a helve. The beam was attached to the ram by leather straps, and a rubber cushion was fitted between the beam and the top of the ram. (Bradley Company trade literature ca. 1935)

Deering's next patents, numbers 312,826 and 355,146 which were granted on February 24, 1885, and December 28, 1886, pertained to the Upright Helve design. The first of these patents detailed the guided-ram, the means of connecting the ram to the helve, a sliding friction sleeve on the pitman (which permitted adjustment of the stroke), and the new mechanism for communicating motion to the helve. The second patent focused on the treadle, brake and idler pulley. The ram of the Upright Helve was connected to the beam by two leather strips on either side of a rubber cushion. Several Upright Helve Hammers were in operation as early as 1884, but production began in earnest the next year.[30] This design was so robust that Bradley was able to offer the machine in 300- and 500-pound models. It was recommended for service which was continuous and severe.[31]

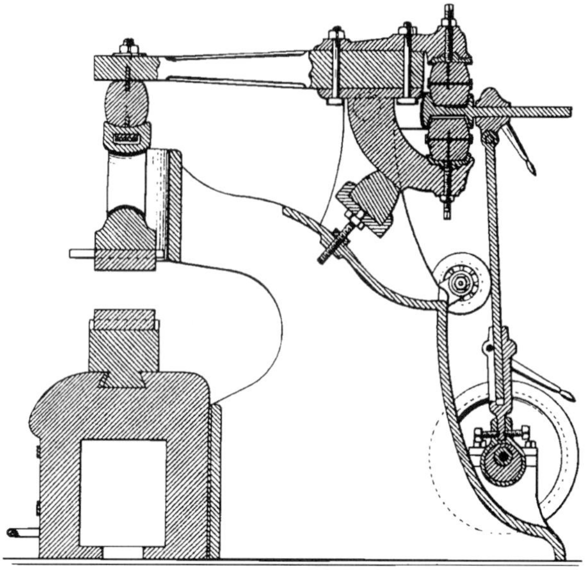

The Upright Helve Hammer was described in Michael Deering's patent of 1885.

The design of the Upright Strap Hammer is outlined in Deering's two patents of 1887, numbers 355,512 and 368,384 which were granted on January 4 and August 16. Both patents focused on the method of suspending the strap, with the second patent concerned largely with the means for adjusting the strap's tension (it was found that tightly stretched straps were slower to overheat).[32] The configuration of the "helve" which is depicted in the

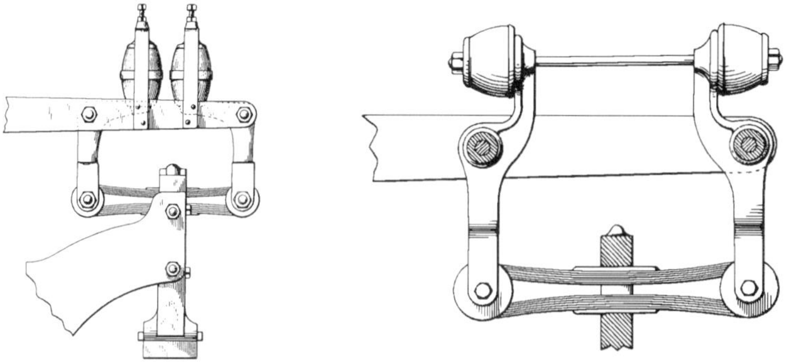

These two variations of the rubber-cushioned carrier for the leather straps were described in Michael Deering's 1887 patent #355,512.

The Bradley Upright Strap Hammer was designed for intermittent work requiring rapid, snappy blows - such as the manufacture of cutlery. (*The Iron Age*)

drawings of the second 1887 patent (built-up from steel strips with wood in-between) is very much like that which was introduced in *The Iron Age* in February of 1890.[33] Before the turn of the century, the beam was redesigned, eliminating the wood section and positioning the steel strips as the sides of the "helve." This configuration is shown in the drawings of Frank Raymond's patent number 575,784 which was granted on January 26, 1897, and was assigned to Bradley. This patent also dealt with the continuing problem of strap tension and overheating. The Upright Strap Hammer used a well-established means of suspending a forge hammer ram. The leather strap increased the possible throw of the ram on the Bradley hammer, and imparted a snappier blow. This also increased the range of work which could be done with the

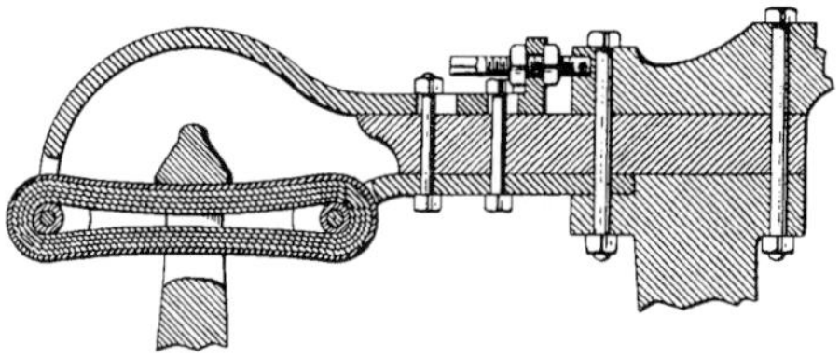

Above, Michael Deering's patent from 1887 describes the early style of the Upright Strap Hammer, with its steel and wood beam. Below, Frank Raymond's patent of 1897 describes the all-steel beam, and another means of tensioning the straps.

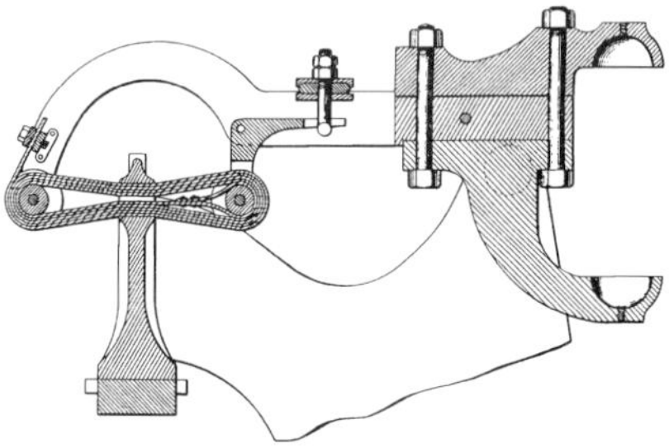

hammer, but the leather strap introduced a limitation. Soon after these machines were placed on the market in the late 1880's, it was found that the leather tended to overheat quickly (and fail) in the larger sizes of hammers. The 300- and 500-pound strap hammers were quickly discontinued, and that machine was recommended for work of an intermittent nature, with frequent stops between operations.[34]

There were two additional patents relating to the Upright Strap Hammer design, both granted on May 9, 1893. The first, number 497,012 was granted to C. C. Bradley himself, and was primarily concerned with a spherical ball-joint used to connect the pitman to the oscillating carriage of the Rubber Cushioned Helve Hammer. Interestingly, in the patent drawings, this ball-joint detail is also illustrated on a strap hammer of the type depicted in Deering's patent of 1887 (with the rubber cushions oriented horizontally). Hammers of this type were indeed built. An actual example can be seen in the background of a photograph of the hammer shop of the Green River Works of the John Russell Cutlery Company.[35] The second patent, number 497,013 was granted to

A Bradley Upright Strap Hammer with horizontally-oriented rubber cushions appears (behind two Beaudry hammers) in this photograph of the Green River Works of the John Russell Cutlery Company. (Photograph courtesy of the Russell-Harrington Cutlery Company)

Chicago inventor Charles M. Brown, and assigned to Bradley. It depicts a variation of the guided-ram strap hammer with a novel arrangement of the rubber cushions (as well as an adjustable anvil and adjustable guides for the ram). An important part of this patent deals with the ongoing problem of tensioning the leather strap. Previously, Brown had designed a hammer that once offered stiff

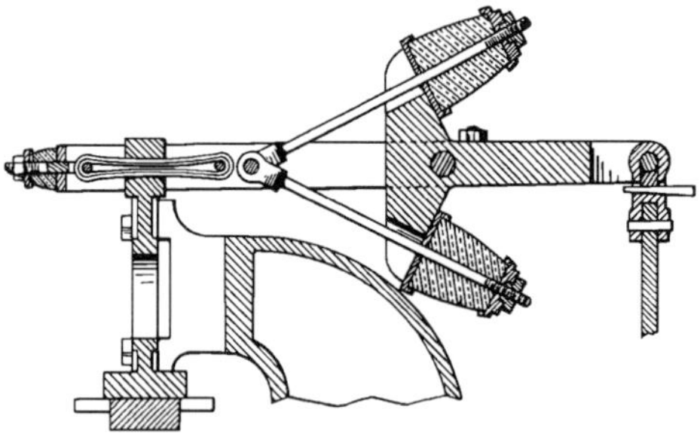

Charles Brown considered the problem of tensioning the leather straps of the Bradley hammer in his patent of 1893.

competition to the Bradley Helve during the 1880's: the Brown Slow Raising, Quick Striking Hammer, built by the Knowlton Company of Rockford, Illinois.

The non-vertically configured Bradley hammers were manufactured for more than half a century alongside the original helve and the vertically-configured Compact Hammer. The Rubber Cushioned Helve was always the flagship of the line, the most substantially-built, and the most expensive Bradley hammer. In the Fall of 1939, for example, a 100-pound Rubber Cushioned Helve (with an electric motor), weighing approximately 6,650 pounds, cost $1455. The 100-pound Upright Helve Hammer and the Upright Strap Hammer (similarly equipped with electric motors) weighed 4,700 pounds and cost $1,105 each. Lighter still, at 3,500 pounds, and cheaper to produce, a comparable Compact Hammer sold for $890. Despite the difficulties encountered in developing the Upright Strap Hammer (regarding tension and overheating of the straps), this machine and the Upright Helve were significantly less complex than the Rubber Cushioned Helve (with its oscillating carriage), and were thus far cheaper to build.[36]

* * *

John C. Butterfield of Chicago, Illinois, was one of the most prolific inventors in the realm of power forging hammers. In 1870, he and James Hay patented the mechanism which soon became the

Bradley Rubber Cushioned Helve Hammer. Butterfield's patents of 1873 reveal that he remained involved in the practical development of the Bradley hammer, though the exact nature of his relationship with the Syracuse manufacturer is not known. He did, however, subsequently pursue a different direction in hammer design, and was granted more than half a dozen additional hammer-related patents after 1873. The author has been unable to locate any information on actual hammers (built and sold) of Butterfield's later design, but the patent evidence strongly suggests that this work was more than an inventor's flight of fancy. These later patents are primarily concerned with atmospheric, or air-cylinder spring, versions of non-vertically-configured guided-ram hammers. However, Butterfield continued to explore the further use of rubber cushions in hammer design.

On October 27, 1874, Butterfield was granted patent number 156,276 for his application of rubber cushions in what he described as "that class of hammers known as upright or dead-stroke power hammers."[37] In lieu of a steel bow-spring, Butterfield interposed two rubber cushions between the crank and the ram. He acknowledged that this approach might also work for a hammer with a pivoting beam, and illustrated the mechanism both ways in the patent drawings. This configuration of rubber cushions is

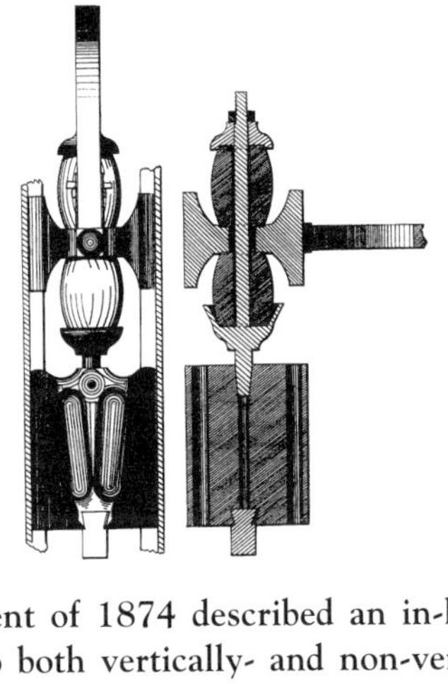

John Butterfield's patent of 1874 described an in-line use of rubber cushions, as applied to both vertically- and non-vertically-configured hammers.

understandable given Butterfield's previous work, but it is unlikely that such an arrangement could provide adequate movement to effectively compete with a bow-spring linkage. In any event, the Bradley Company did not use this linkage design; and when it ultimately brought out the Compact Hammer, its rubber-cushioned linkage was quite different.

Butterfield's in-line rubber-cushioned elastic linkage did figure in his own hammer designs. On April 18, 1876, he was granted patent number 176,400 for a most unusual variation on the atmospheric hammer theme. In that patent, Butterfield pointed out that air-cylinder linkages were imperfect in protecting the machine from the transmission of shock (since vibration was transmitted through the close packings which were required to prevent the escape of air under pressure). To eliminate this problem, Butterfield

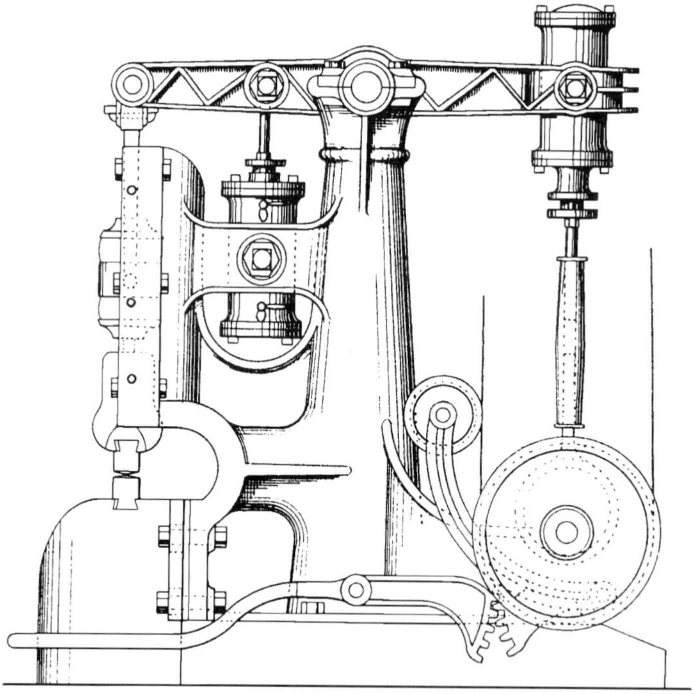

John Butterfield's patent of 1876 described a most unusual version of the atmospheric hammer: the air-cylinders regulated the hammer while the previously described rubber linkage provided cushioning.

used his previously-patented rubber cushioned linkage to block the transmission of shock. The two air cylinders used in Butterfield's design were used solely to operate and control the action of the hammer (and not to provide insulation from shock as in previous atmospheric hammers). The supplemental air cylinder, which was located between the pivot and the ram, used valves to regulate the hammer's stroke. This allowed the stroke to be varied without changing the hammer's speed. This was not possible with the previous atmospheric hammers on the market. Butterfield also pointed-out that vertically-configured atmospheric hammers (like Hotchkiss' hammer) required enough power to not only overcome friction and inertia, but also to lift the dead-weight of the hammer ram. Butterfield's hammer, with its pivoting beam, was claimed to

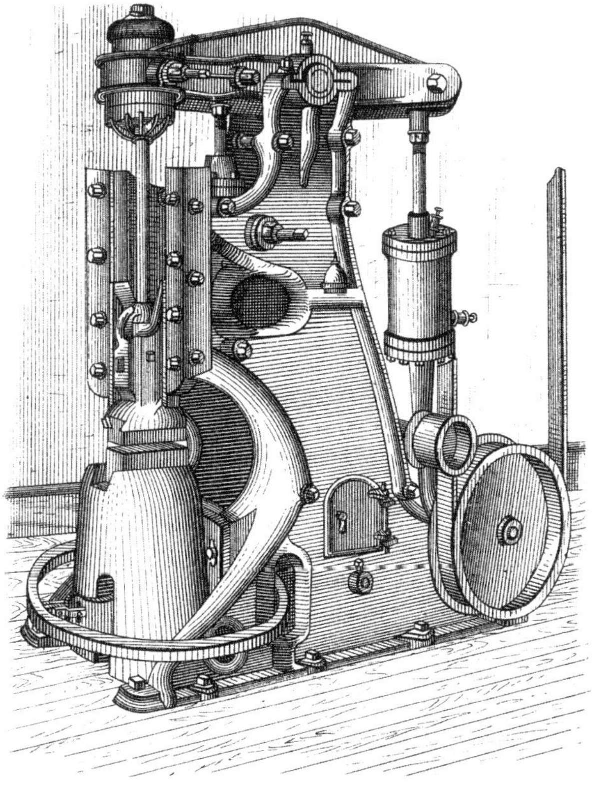

An improved version of John Butterfield's atmospheric hammer appeared in his patent of 1882.

be counterbalanced, requiring a minimum of power for its operation. Toward this end, the design also included a relatively light cast-iron beam with a distinctive webbing for strength. Additionally, Butterfield's patent included a novel treadle and idler-pulley arrangement.

A refined and very real-looking version of Butterfield's hammer is illustrated in his next patent, number 264,859 which was granted on September 26, 1882. This patent described 1) an improved connection between the hammer ram and the rod, 2) the placement of the tightener-lever centrally within the machine, 3) a brake which acted upon the hammer ram, 4) an adjustable cylinder head to regulate the length of the stroke, and 5) a wrought metal beam formed from two parallel bars. It is interesting to observe that the frame of the hammer was to be manufactured in two halves (left and right sides) which were bolted together (an unusual design which might have inspired Alexander Beaudry in his thinking). Additional refinements were patented in the next year. Patents number 285,387; 285,389; and 285,390 were all granted to John C. Butterfield on September 25, 1883. The first of these described the use of a rubber damper to reduce vibration and fatigue of the steel hammer rod. The second patent described a

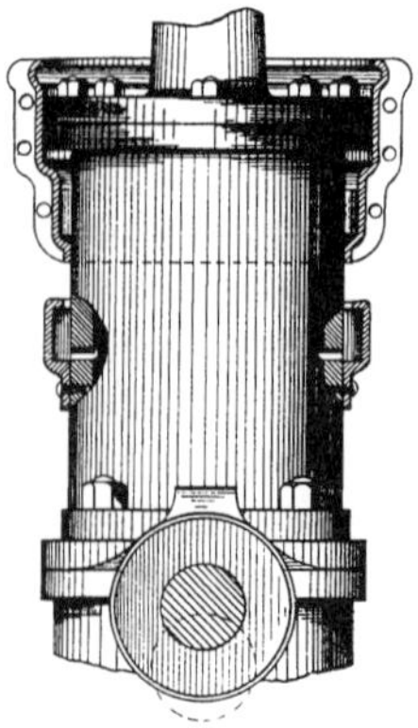

Left to right, John Butterfield's patents of 1883: number 285,387 for a rubber damper, number 285,389 for a coil-spring cushioned pitman, and number 285,390 for a cylinder-cooling water-jacket.

means of protecting from strain the bearings at the rear of Butterfield's hammer's beam (by means of coil springs). The third patent described a jacket for water-cooling the rear air-cylinder. These practical improvements leave little doubt that there were at least prototype models of this machine in use. A fourth patent from 1883, number 290,850 which was granted on Christmas Day, describes a lubrication system for the air cylinders "employed on my atmospheric hammer and stamp-mill." This suggests the possibility that Butterfield's designs were realized primarily in a non-forging application.

* * *

William Rhys Jenkins, Jr. of Bellefonte, Pennsylvania, was a contemporary of inventors Beaudry, Deering, and Butterfield. Hammers embodying Jenkin's design appeared at about the same time as the Beaudry and Bradley Uprights, during the early years of the prosperous 1880's. Known at first as the Vulcan or the Square-Stroke Power Hammer, and later as Jenkin's Cushioned Power Hammer, these machines held a place in the power hammer market for more than two decades.

Jenkins was born in Wales in 1844. He came to the United States with his parents in 1852 and settled in Howard, Pennsylvania. His father secured an interest in the Howard Iron Works, and young Jenkins learned about iron and its manufacture by working in the rolling mill as soon as he became old enough to do so. He apprenticed to learn the machinist's trade in the shops of Todd and Duncan in Bellefonte, and went on to work in several other large plants in Milton and Williamsport. In 1879, Jenkins returned to Bellefonte; and, with J. H. Lingle, bought-out Todd's interest in the firm. The new partnership became known as W. P. Duncan and Company.[38] Several years after this change, in 1883, the company was described as occupying one and a half acres, with power provided by a water wheel of twenty-five horsepower. About thirty men were employed in its foundry, machine, and black-smithing shops. Four hundred tons of iron were reportedly used annually in the manufacture of stationary steam engines, turbine water wheels, ore washing and mining machinery, and square-stroke power hammers.[39]

Jenkins' hammer ca. 1888. (from "Notes on Steam Hammers," by C. Chomienne in *The Railroad and Engineering Journal*)

The hammer built by W. P. Duncan and Company featured a rubber-cushioned fulcrum for the helve, and a guided ram provided the touted "square" stroke. The design of the fulcrum first appeared in patent number 237,872 which was granted to William R. Jenkins, Jr. on February 15, 1881. In what was perhaps the simplest design for a rubber-cushioned hammer, a forged steel helve rocked upon a pivot which was held (on either side of the beam) between rubber springs. This design was straightforward and effective, and the hammer was also relatively inexpensive to build. In 1885, the hammer was offered in 60-, 80-, 100-, and 125-pound sizes for $375,

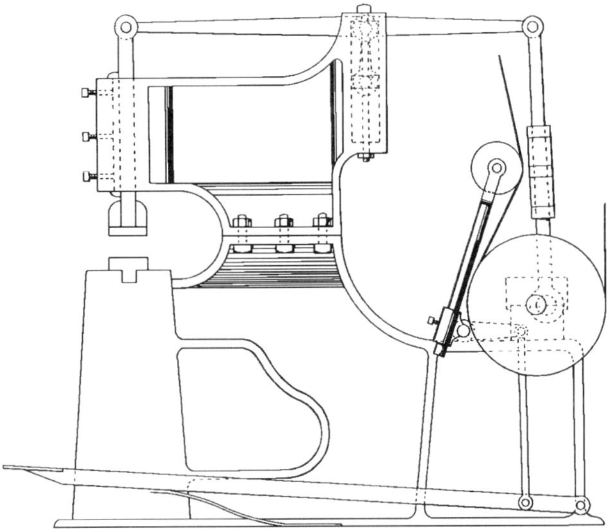

William Jenkins' patent of 1881 described his rubber-cushioned pivot.

$400, $450, and $600, respectively, (other sizes were available by special order).[40] A comparable 100-pound Bradley Upright Cushioned Strap Hammer cost $600, and a Rubber Cushioned Helve cost $850.[41] By the early 1890's, the line had been expanded to include lighter hammers of 25, 40 and 50 pounds, as well as heavier 150- and 200-pound models.[42]

Trade literature from the company made a point of stating that it had "reduced the principle of cushioning to its utmost simplicity," and that construction of its hammer was "simple and tasteful" throughout.[43] Jenkins' subsequent hammer patents respected the virtue of that simplicity and described only subtle improvements. In patent number 358,672 which was granted on March 1, 1887, Jenkins proposed an interesting variation of the cushioned fulcrum which employed two elliptical rubber springs in lieu of the four rubber blocks. He also included a similarly-configured connection between the helve and the ram, which was intended to add greater elasticity to the blow. A second patent of 1887, number 370,260 was granted on September 20. It revealed another version of the pivot which elevated the fulcrum by securing

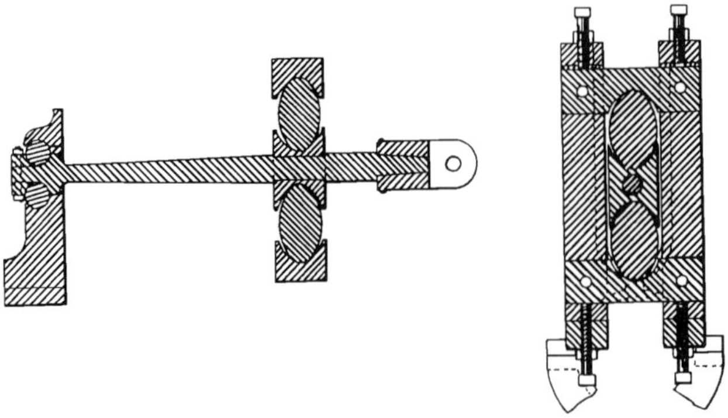

William Jenkins patented several improvements in 1887: number 358,672 (above) described a new configuration of the rubber cushions; number 370,260 (below) described an alternative suspension for the beam, and a circular arrangement to hold the straps from which the ram was suspended.

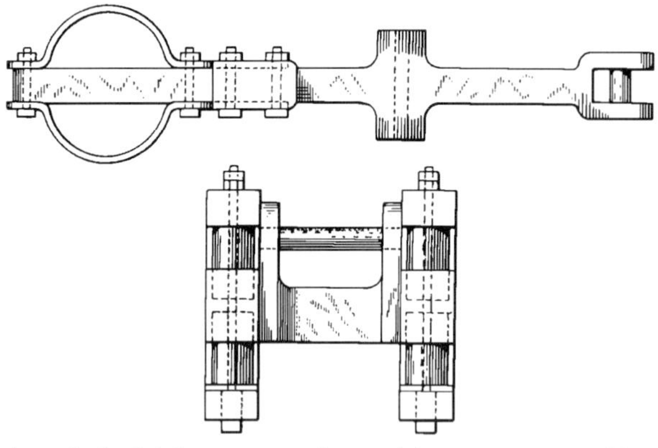

it in a cross-head, held between the rubber springs. Furthermore, the ram was suspended from a leather strap which was held by a horizontally-oriented circle formed by two flat spring-steel bars. While it is not known if the improvements from the first patent were ever incorporated into actual machines, hammers with the strap linkage were indeed built for several years. In his article on steam hammers, which appeared in 1888-1889, French engineer C. Chomienne illustrated the "Duncan Helve Hammer," (sic) describing the aforementioned strap linkage. He pronounced the hammer "an exceedingly useful tool for small forgings and die work."[44]

Jenkins' next patent, number 466,790 granted on January 12, 1892, proposed a significant modification to the connection between the helve and the ram. Instead of a simple, hard connection (as used in the earliest Vulcan hammers) or an elastic linkage (as described in his 1887 patents), Jenkins merely placed the end of the wood helve through an opening in the ram, allowing

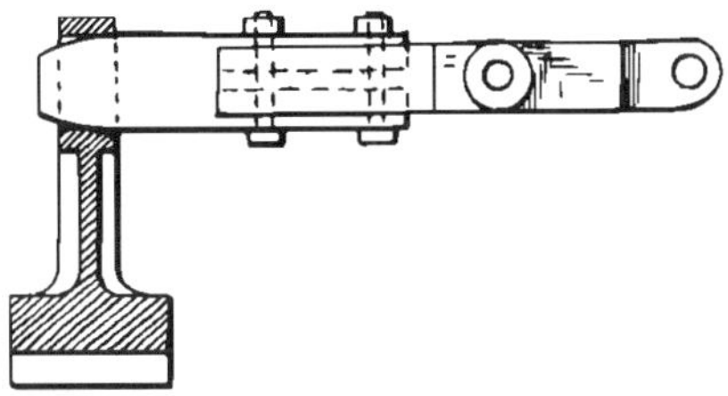

William Jenkins' patent of 1892 described a loose connection between the ram and the end of the beam.

it to fit loosely. A hammer which featured this modification, as well as the fulcrum-raising crosshead design, was described in a profile of the hammer in *The Iron Age* in March of 1891. The reviewer expressed surprise that a loose, non-cushioned linkage of the helve and ram had not appeared earlier. The builders claimed that the use of steel springs and leather straps to secure the ram to the helve was, in fact, the weak point in previous upright hammer designs.[45]

In 1888, W. P. Duncan retired, and the company became Jenkins and Lingle. Production of the hammers continued and even included an export to W. A. Sanberg of Sweden.[46] Jenkins retired from the company in 1902, returning to Howard where he and his brother organized the Howard Iron and Tool Company. Jenkins' last hammer-related patent was number 736,971 which was granted on August 25, 1903. It described a machine intended primarily for chain-making. Surprisingly, this hammer was a trip-actuated helve, with an air-cushion to regulate the stroke and a "catch-up," or trigger, to stop the helve in the upper position. This anachronistic design was probably never manufactured. William Rhys Jenkins, Jr. died in 1908. He was remembered in Bellefonte for his power hammer, which was hailed as being "universally known among iron manufacturers," as well as the inventions of a continuous rolling mill, water wheel, low water alarm, coal burner and machinery for mass-producing hand tools.[47]

Following Jenkins' retirement in 1902, the Cushioned Power Hammer had remained the property of J. H. Lingle. He continued to build it for several more years. On November 15, 1904, patent number 774,791 was granted to James D. Seibert of Bellefonte, with half the rights assigned to J. H. Lingle. The patent describes another slight variation of the cushioned pivot as well as an adjustable crank and a double coil spring linkage to cushion the pitman. This latter detail is highly reminiscent of the early designs patented by Alexander Beaudry. The extent to which these

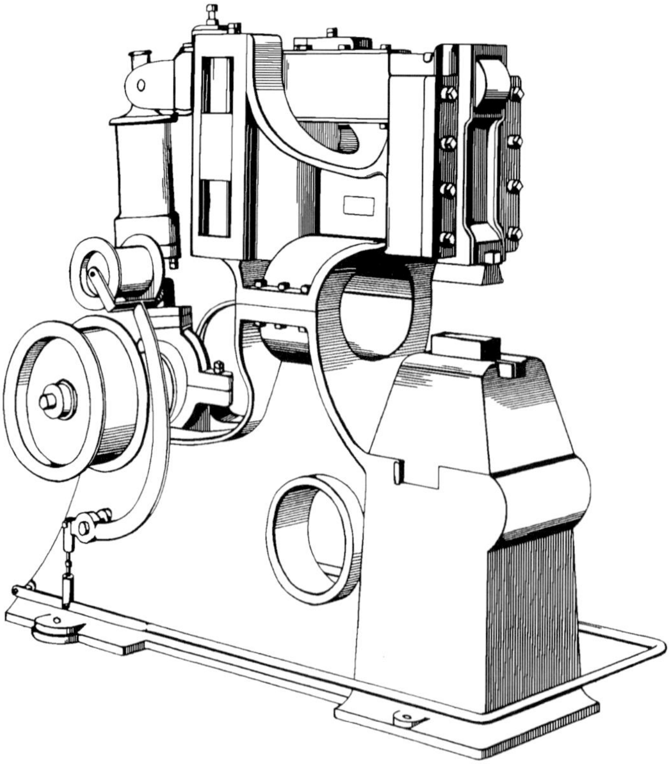

James Seibert's improvements to Jenkins' hammer were patented in 1904, after William Jenkins' retirement.

modifications were incorporated into J. H. Lingle's hammers is not known; and after 1907, the company was no longer listed in the *Thomas' Register of American Manufacturers*.

* * *

Arthur Alfred Koch filed for a U.S. patent on the salient features of his hammer in February of 1903; and on February 9, 1904, he was granted patent number 751,522. His Modern Power Hammer employed a pivoting beam with a movable fulcrum to provide variable blows. The beam itself was constrained from front-to-back movement by steel straps, while the pivot shaft could be moved forward and backward (within the beam) by a long lever with detents to hold it in the desired position. Steel bearing-blocks on either side of the pivot moved within inclined brackets. This brought the ram closer to the anvil when lighter blows were delivered or farther away for heavy blows. A bow-shaped steel spring was interposed between the ram and the beam, both to store energy for a forceful blow and to isolate the machine from the shock

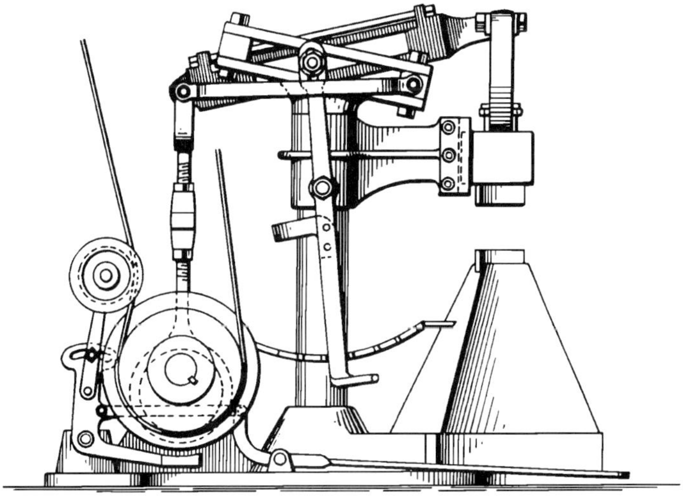

Arthur Alfred Koch's movable-fulcrum hammer was patented in 1904. Hammers with the long shift-lever (as shown in the patent) were manufactured in Montezuma, Iowa, and very briefly in Grinnell.

of impact. An adjustable connecting-rod linked the beam to an eccentric (a disc which was keyed, off-center, to a straight shaft). This rod could be shortened or lengthened to accommodate gross changes in the thickness of the work. The operation of the hammer was controlled by a foot-treadle which tightened an idler pulley against the drive belt when depressed (enabling some modulation of the rate at which the blows were delivered). When the treadle was released, a brake made contact with the wide inner rim of the drive pulley, stopping the hammer. Subscribers to *The Blacksmith & Wheelwright* were told that this was "just the thing for plow work," and assured that it would not go to pieces! At $100 each, Koch was willing to ship his hammers on trial to responsible parties.[48]

The Modern Power Hammer of 1903. Note the unusual orientation of the bow-spring. (*The Blacksmith & Wheelwright*)

The idea of a hammer with a movable fulcrum was technically not new. Daniel Longworth, of Birmingham, England, had developed a movable fulcrum hammer in 1876.[49] That machine was initially manufactured by the engineering firm of J.

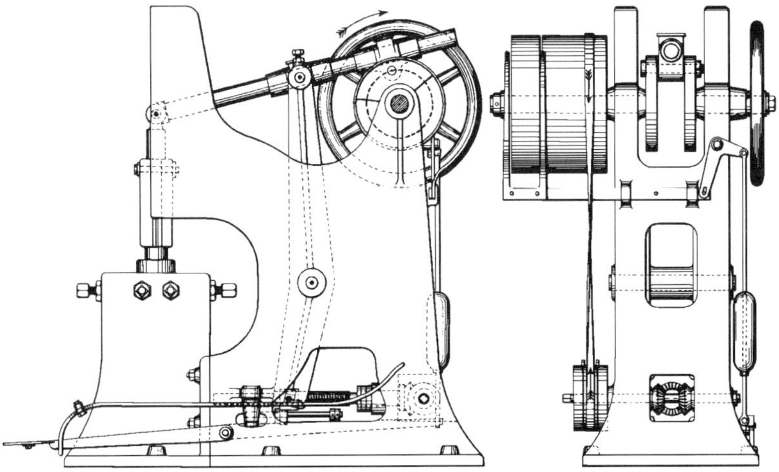

Longworth's movable-fulcrum hammer first appeared in 1876. In this slightly later version, note the belt-powered mechanism for changing the fulcrum by foot-control. (*Proceedings*, Institution of Mechanical Engineers, London)

Scott Rawlings, of which Longworth was the manager.[50] It used an atmospheric, or air-cylinder spring linkage to isolate the mechanism from vibration, and was built in a wide range of sizes. In a paper read before the Institution of Mechanical Engineers (London) in April of 1882, Longworth described the application of his design to both a light plannishing hammer and a 1 cwt forging hammer. The forging hammer was particularly sophisticated, with power-assisted adjustment of the fulcrum: the smith used a foot control to shift a small leather belt to pulleys which moved the fulcrum either forward or backward.[51] Less sophisticated versions of the Longworth hammer were later built, with the fulcrum adjusted by simply placing a pivot pin in one of several holes.[52] The firm of Samuelson and Company of Banbury (which built the Longworth hammer during the 1890's) offered machines with rams weighing from ten pounds up to ten cwt.[53] In 1900, Ernest Samuelson read a paper to the same Institution of Mechanical Engineers describing "Improvements in the Longworth Power Hammer."[54] While Longworth's hammer was thus long-lived and evidently quite successful in England, it does not appear to have made an impression on the American hammer market. Only one contemporary American-built hammer reflects the influence of Longworth's design: Penny's Spring Cushioned Helve Hammer. This curious machine, which was patented in 1884, has already been described in Chapter 3. Though fundamentally a helve hammer, Issac Penny's machine used a pivoting beam and bow-spring linkage to impart motion to the helve. Of significance here is the fact that the fulcrum of the beam could be adjusted to increase or diminish the motion. A lever made it possible to readily move the beam forward or backward within a fixed pivot assembly. The hammer designed by Koch twenty years later placed the pivot itself in motion and gave American blacksmiths a simplified and affordable version of the movable fulcrum hammer.

Koch was born in Missouri on May 10, 1871. The son of a blacksmith, Koch followed in his father's footsteps; but he possessed a mechanical aptitude and ambition that ultimately led beyond the village smithy. As a young man, he moved to Montezuma, Iowa, where his own blacksmithing business expanded to include a wide range of repair services and manufacturing. Koch offered to build "anything and everything to order," and could repair engines, boilers, threshing machines, binders, mowers, wagons, plows, buggies, harrows and even automobiles. Situated in the Monte Foundry and Machine Shops, the Koch Manufacturing Company produced such foundry staples as blacksmith's swage blocks (12" x 12" x 4") which sold for $3.[55] Koch developed a line of blacksmith's and wheelwright's machines which he dubbed the "Modern" line. Included were a tenon & boring machine (for wooden wheel hubs) and a combined punch and shear. The Modern Power Hammer was Koch's most ambitious product, revealing his skill as a mechanic and patternmaker. It quickly attracted the attention of both blacksmiths and enterprising businessmen in cities beyond Montezuma.

Montezuma was predominantly an agricultural town, and Koch began to look for a larger city to support his foundry and manufacturing business. The Koch Manufacturing Company was an attractive proposition and several cities vied for its presence. In June of 1903, the businessmen of Mason City, Iowa, appeared to have the upper hand; but the Commercial Exchange of nearby Grinnell bettered their offer. Several of Grinnell's leading businessmen entered into a partnership with Koch, the citizens of Grinnell furnished land and a building at a cost of $10,000 (to be rented to the company at a nominal rate), and the city council agreed to abate all city taxes on the enterprise for a period of five years.[56] The partnership looked most promising. One of Koch's two partners was Henry W. Spaulding, a one-time blacksmith who was already well-established in Grinnell building carriages. As president of the Commercial Exchange, Spaulding worked tirelessly for Grinnell, and ultimately served two terms each as Mayor and State Senator.[57] Koch's other partner was J. C. Goodrich, whose management skills had been honed by running the town's water works. With such an organization, the erection of the new buildings and the foundry progressed rapidly.

On November 27th, 1903, the whistle of the Koch Manufacturing Company joined the morning concert of whistles from Grinnell's factories.[58] With a crew of men still at work installing line shafts and machinery, the first run of castings for power hammers was produced in early December. A backlog of orders for the hammers prompted an initial schedule of three castings per week;[59] but with a dozen hammers completed and eight of them sold, it was decided to halt production and redesign the machine. The redesigned hammer was substantially heavier, more robust, possessed a greater range of stroke and an improved guide for the ram. Also, the lever used to shift the fulcrum was shortened and a simplified bow-spring arrangement was devised. The first hammer of the new design was shipped to the Iowa Manufacturing Company of Oskaloosa in mid-February of 1904. The Koch Manufacturing Company was able to construct one hammer per day and expected that rate to increase. Built in a single size with a 35-pound ram, the hammer now weighed 1,200 pounds overall and cost $150.[60] H. W. Spaulding was in charge of sales, and announced that he expected to sell them by the trainload. Indeed, an important factor in Koch's decision to move to Grinnell was undoubtedly its prime location at the confluence of the Rock Island

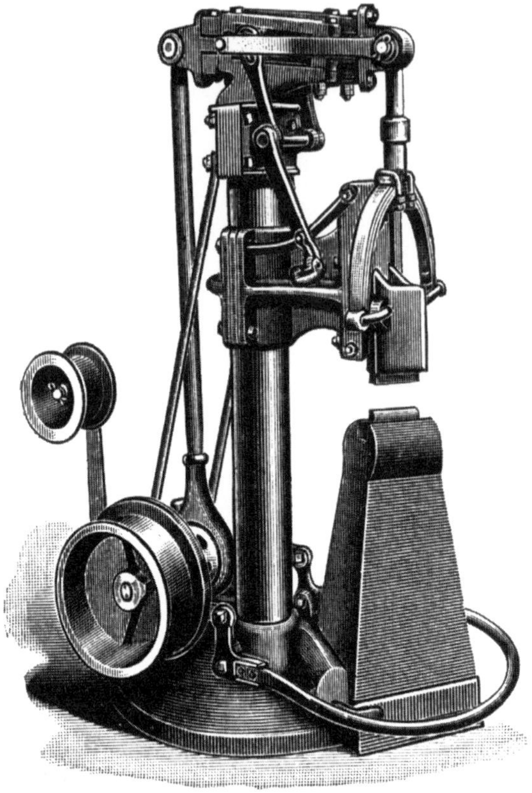

Koch's Modern Power Hammer was revised in 1904. Note the shortened shift lever and the more-conventional orientation of the bow-spring linkage. (*The Blacksmith & Wheelwright*)

and Iowa Central Railways. As superintendent of the new foundry and machine shop, Koch oversaw a workforce of about two dozen men. Some of his employees had moved with him from Montezuma, and they would move with him again when he left Grinnell. Unfortunately, that proved to be all too soon.

On July 5, 1904, it was quietly announced in the *Grinnell Herald* that Koch had sold his one-third share in the company along with the patent rights to the power hammer.[61] The *Mason City Globe Gazette* reported that some difficulty had arisen between Koch and his partners in Grinnell, but there was no elaboration.[62] Koch soon purchased the residence and foundry of Henry Hathorn

Arthur Alfred Koch (center) and his crew shortly after leaving Grinnell (ca. 1906). (Photograph courtesy of Darlene Koch Deason)

in Mason City, and began production of the vertically-configured Vulcan Power Hammer (this hammer and Koch's subsequent activities are discussed at length in Chapter 4).

The Grinnell Manufacturing Company continued to both manufacture and develop the Modern Power Hammer after Koch's departure. John W. Fitzgerald assumed Koch's interest in the company and was placed in charge of sales. In September of 1905, a second revision of the hammer design appeared in advertisements in *The Blacksmith and Wheelwright*. The shifting lever was repositioned, pointing upward, with the detents located along the top of the pivot brackets. The base casting was also redesigned to include the lower drive-shaft bearings as one unit. In the Spring of 1908, the Grinnell Manufacturing Company was reorganized. Spaulding withdrew from active involvement in order to concentrate on the retail buggy business.[63] The Grinnell

This advertisement for the Modern Power Hammer appeared shortly after A. A. Koch's departure. (*The Blacksmith & Wheelwright*)

Manufacturing Company shop closed down briefly, but re-opened in mid-August. With a new foreman, William Butler, at the helm, the shop concentrated on its general repair business.[64]

Soon a new hot product appeared in Grinnell: electric washing machines! The Thompson Brothers had moved their enterprise (and patented products) from Des Moines, Iowa, and many of Grinnell's leading businessmen were eager to join the venture. J. C. Goodrich became manager, and washing machines were soon being assembled in the shops of the Grinnell Manufacturing Company.[65] In July of 1909, the Speed Sales Company was formed to handle sales and distribution of Grinnell-built washing machines, the Speed and the Elmo. The manager of this company was E. E. Clock, and he shared office space with the Grinnell Manufacturing Company.[66] In October, the Thompson Brothers Company announced that "owing to the growth of our business we are obliged to discontinue all job and repair work."[67]

Meanwhile, manufacture of the Modern Power Hammer had continued in Grinnell (with distribution handled by the Modern Sales Company since the Spring of 1908). Proclaiming its hammer "the backbone of blacksmithing," the machine was little changed from the 1905 version. In August of 1910, it was reported that E. E. Clock had returned to his former home in Hampton, Iowa, "where he will assist his brothers in the abstract business and continue the management of the power hammer which he is now pushing."[68] Advertisements from the Modern Sales Company of Hampton

The Modern Power Hammer was again revised, with the shift lever re-oriented upward. In 1910, the hammer was being sold out of Hampton, Iowa. (*The Blacksmith & Wheelwright*)

noted that it was licensed by the Grinnell Manufacturing Company, and the line continued to evolve. Advertisements in *The American Blacksmith* touted "the new square base... a marked improvement, adding about one hundred pounds to the weight of the machine."[69] The hammer was now offered in four styles and three sizes: 35, 50 and 75 pounds. One of these new styles, named the "Little Iowa," appeared in an advertisement in *The Blacksmith and Wheelwright* of December, 1913. Only faintly recognizable as a descendant of Koch's Modern Power Hammer, the fulcrum was no longer adjustable. At $75, though, a Little Iowa No. 3 with a 35-pound ram was among the lowest-priced hammers in America.[70] The full line of Modern Power Hammers was described in a Canadian trade catalogue from this period. D. Ackland & Son of Winnipeg, Manitoba, offered the Little Iowa as its Little Modern, and a 50-pound version of the non-moveable fulcrum hammer was known as the New Modern. The square based moveable-fulcrum Modern Power Hammer was offered in two sizes: 50 and 75 pounds.[71] After 1914, the Modern Sales Company was no longer listed in the *Thomas' Register of American Manufacturers*.

*   *   *

Hammer designer Henry Alinder of Milwaukee, Wisconsin, adopted Jenkins' method of loosely linking the helve to the ram without an elastic linkage. Alinder had emigrated to the United States from Sweden, where he may have gained experience with one of Jenkins' exported Cushioned Helve Hammers. In 1908, he applied for a U. S. patent on his own hammer, prior to becoming an American citizen. Eventually, Alinder was granted patent number 966,412 (on August 9, 1910) for what he claimed was a "milestone in the art of power hammer design and performance."[72]

The Alinder Hammer. (Photograph courtesy of the Chambersburg Engineering Company)

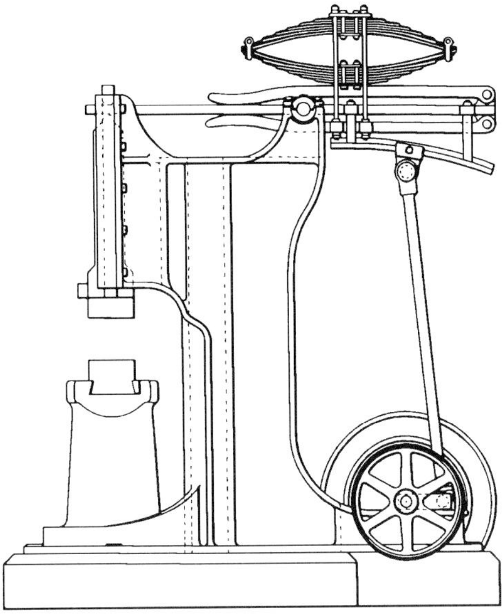

Henry Alinder's patent of 1910 described a sophisticated arrangement of spring-compressed levers which provided the "Alinder action."

Alinder's hammer employed one of the most sophisticated spring-cushioning arrangements to be devised. A vanadium steel helve was captured between two levers which were held together by a substantial, double leaf-spring mounted above. The spring stored unused energy at both the beginning and the end of the stroke, giving "power, speed, elasticity and wonderful life to the blow."[73] While this "Alinder action" could accomodate a wide range of thicknesses without requiring adjustment, the stroke could be quickly varied by moving the upper end of the pitman along a distinctive curved bracket. The connection of the pitman was thereby brought nearer to, or farther from, the fulcrum of the beam. A heavy flywheel on the drive shaft absorbed the energy returned by the spring, and insured steady and even operation. Alinder claimed that any "dancing or vibration of the hammer head is absolutely avoided."[74]

Manufacture of the hammer was soon under way. Initially it was offered by the Reliance Engineering and Equipment Company of Milwaukee, who described the Alinder hammer though unnamed) in a notice that appeared in *Machinery* in January of 1911.[75] This company was headed by consulting engineers Claude A. Tupper and Frank G. Booles, but otherwise little is known about the extent of its operation.[76] By November of 1917, a separate entity, the Alinder Hammer Company, had been formed. The hammer was then available in four sizes: 40, 75, 125 and 250 pounds. The largest (No. 4) hammer was substantial indeed, weighing 5,000 pounds overall. A photograph of this hammer reveals a robust and elegantly-proportioned machine. A drawing of the smallest (No. 1) hammer dated "Feb 8-13" is included in trade literature of the Alinder Hammer Company. It was drawn by Alinder himself and it is known that he also worked as a draftsman for Allis-Chalmers during the period from 1906 through 1918 (though perhaps intermittently).[77] Unfortunately, little is known of the short-lived Alinder Hammer Company, and it never appeared in the *Thomas' Register of American Manufacturers*. There is a strong

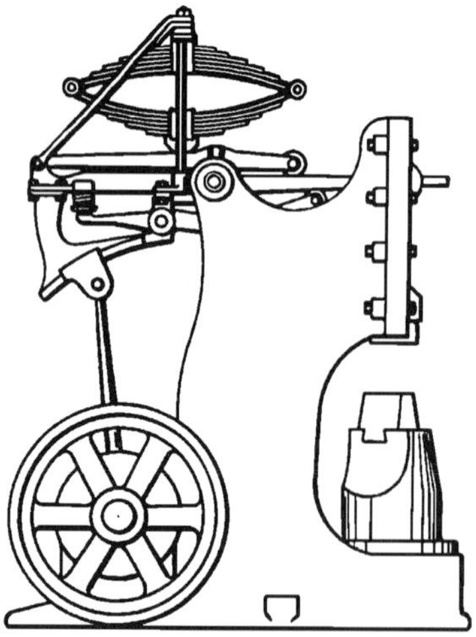

Anders Wallin's patent of 1924 suggests that the Alinder hammer may have emigrated to Sweden. Wallin added rollers to the ends of the levers to diminish friction and wear.

possibility that the hammer (and perhaps the inventor as well) moved to Sweden. On April 29, 1924, Anders Wilhelm Wallin was granted a U. S. patent for his improvement to what is clearly an Alinder machine. Wallin was self-described as a "chief engineer," and the patent is assigned to the Aktiebolaget Klafrestroms Druk, of Klafrestrom, Sweden. The patent described the addition of roller bearings where the spring arms bear on the beam (eliminating friction and wear, and improving efficiency).

* * *

Charles Keller was another Swedish imigrant whose mechanical skills turned briefly to the design of power forging hammers. He was born in 1873, the son of a blacksmith, and came to the United States in 1892.[78] Shortly before the turn of the century, Keller moved to Bloomer, Wisconsin, and established a first-rate machine shop there in the "Old Norsk" Saloon Building.[79]

The playful design of the Comet's frame harkened back to the nineteenth century. (*The Blacksmith & Wheelwright*)

A rare image of a power forging hammer and its builder: Charles Keller with his Comet. The incongruous forest setting was cut away to provide an image for company advertisements. (Photograph courtesy of Tom Enderson). Below, Charles Keller's patented friction clutch.

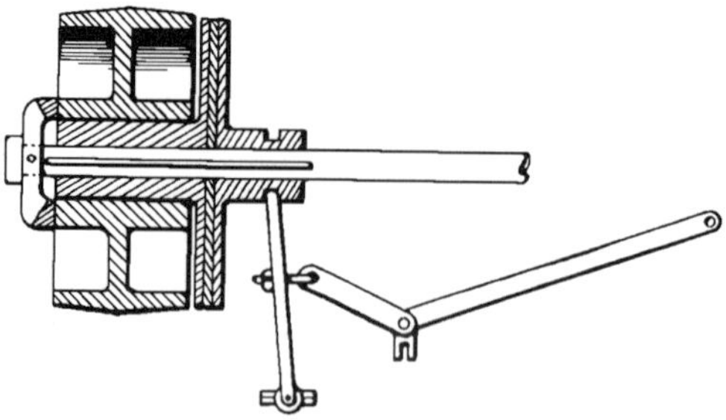

The shop was initially powered by a water wheel, but a steam engine was soon installed. Within a few years that too was replaced by a gasoline engine which ran both the line shafts and a dynamo for electric lighting. A foundry for casting iron was built in 1901. Charles Keller was joined, in 1902, by his elder brother, J. E. Keller, who was a blacksmith. The business became known as the Bloomer Machine Works and a well-equipped blacksmith's shop was added. J. E. Keller remained in Bloomer for only a few years,[80] but it was during those years that the Bloomer Machine Works began to produce two distinctly different power hammers. One of these, the Thor, has already been mentioned in the discussion of crank-actuated helves, but the other, the Comet, was a non-vertically-configured, guided-ram hammer. Charles Keller applied for a patent on the clutch mechanism of the Comet hammer in December of 1903. He was granted patent number 789,219 on May 9, 1905; and the patent drawings illustrate the clutch as it appeared on the Keller brother's distinctive machine. The Comet was mechanically quite simple, but stylistically it was both playful and elegant. A bow-spring linkage connected the ram to a pivoting round rod. A sliding bracket connected the rod to the pitman, allowing adjustment of the stroke. The patented clutch used a friction disc which was keyed to the drive shaft, and which made contact with the loosely-spinning drive pulley when the foot treadle was depressed. This arrangement eliminated the need for an idler pulley, and the belt tension was constant whether or not the hammer was in use. The most striking feature of this hammer, though, was its sweeping, curved frame with graduated, circular openings. While many hammer designers acknowledged the sculptural potential inherent in these machines, few did so as overtly.

The Thor hammer was probably short-lived, but the Comet was manufactured for more than a decade. In 1909, the Bloomer Machine Works began to manufacture the gasoline engines for which it is best remembered today. The company grew dramatically, and during one month in 1914 it was noted that the foundry was turning-out 20 tons of castings for engines and hammers.[81] Charles Keller left the company in 1915, but returned the next year when the engine-building enterprise relocated to Eau Claire, as the Eau Claire Manufacturing Company. While Keller engines continued to be built for several more years, it is unknown if production in Eau Claire ever included hammers. The Bloomer Machine Works continued to operate under different ownership, and it is possible that hammers were built for a brief time. The Bloomer Machine Works were, in fact, listed in the 1917 edition of the *Thomas' Register of American Manufacturers*.[82]

* * *

The hammer designed by Charles Leonhardt of New Ulm, Minnesota, was among the last of the non-vertically-configured hammers with a central fulcrum to remain on the market. Only the Bradley Upright Hammers survived beyond the 1920's. Leonhardt's hammer appeared at about the same time as Charles Keller's Comet. Both designs included a clutch arrangement, but Leonhardt's machine was far more ambitious and sophisticated. The hammer featured a means of conveniently raising and lowering the fulcrum (up to four inches) to accomodate a wide range of thicknesses. The fulcrum bearings were supported atop a screw which could be turned, through bevel gears, by a crank mounted on the right-hand side of the hammer. The hammer's stroke also could easily be adjusted. Instead of the usual crankshaft, Leonardt's hammer used a crankplate with a T-slot, much like many of the vertically-configured hammers on the market. The hammer also employed a counter-balance that held it in the open position between uses. A Dupont-style, coil-spring linkage protected the mechanism from vibration and shock, using distinctive, long links to connect it to the ram. Both the short toggle-arms and the links pivoted to accomodate the arc of the reciprocating beam. The salient features of this hammer were protected by patent number 765,020 which was granted to Charles Leonhardt on July 12, 1904.

Leonardt's hammer was initially built at his New Ulm Foundry and Machine Shop. This foundry had been established in 1870[83] and was purchased by Charles Leonhardt and his brother in

1879.[84] The business successfully specialized in the repair of agricultural machinery and produced such foundry staples as gratings and graveyard fences. In 1896, the foundry burned and only a portion of the machinery was saved.[85] But Leonardt soon rebuilt, and in May of 1903 he applied for his hammer patent. The hammers built at the New Ulm Foundry and Machine Shop were profiled in *American Machinist* in 1907, at which time one model with a 90-pound ram was being built.[86] In March of 1910, Charles Leonhardt sold the New Ulm Foundry and Machine Shop to the Eagle Roller Mill Company for $8,000. However, he retained possession of the patterns and foundry machinery, and soon re-established himself in the foundry business.[87] He continued to

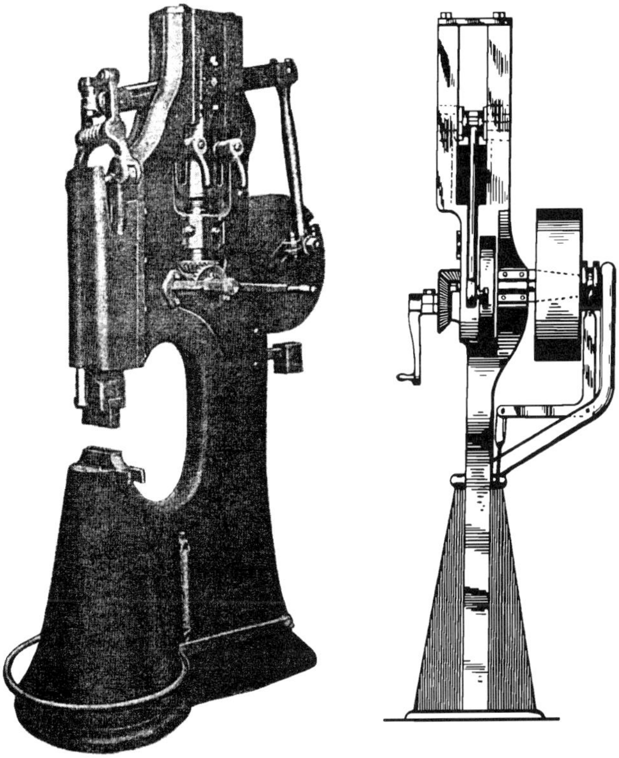

Charles Leonardt's New Ulm Power Hammer was patented in 1904. The patent art shows the mechanism for raising or lowering the pivot, as well as the cone friction clutch. (*American Machinist*)

manufacture his hammer beyond the First World War, and was listed in the *Thomas' Register of American Manufacturers* through the 1923-1924 edition.[88]

* * *

The earliest non-vertically-configured hammer to have its fulcrum at the end of the beam was the machine designed in the late 1880's by Charles A. Vaughn of Newark, New Jersey. He was granted patent number 412,809 on October 15, 1889, for a hammer with an "adjustably regulated... length of drop," said to be of simplified construction and reduced cost of manufacture.[89] In Vaughn's design, the ram was suspended from a leather strap which was held by a semi-circular steel spring (much like the arrangement used earlier in Bradley's early Upright Strap Hammer). The semi-circular spring extended from a forked helve which was reciprocated by a long connecting rod (pitman). The connection point of the

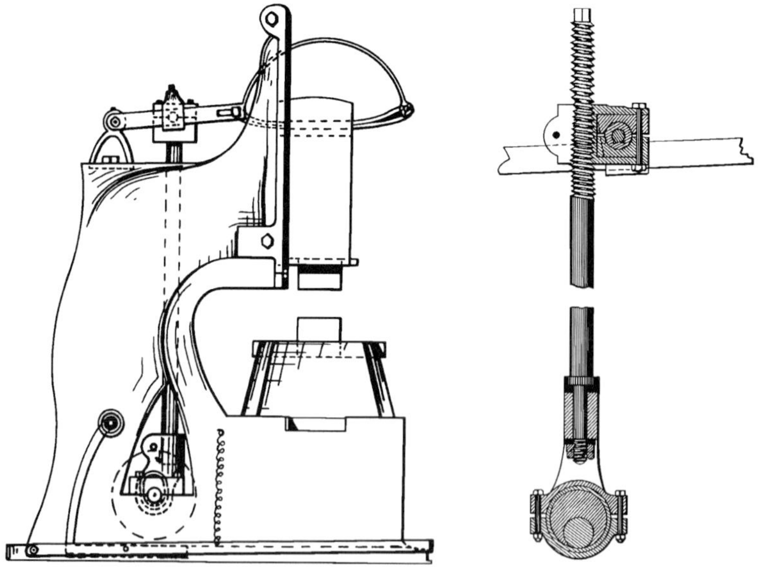

Charles Vaughn's design for a hammer similar to Bradley's Upright Strap Hammer was patented in 1889. In 1891, Vaughn patented an improved, cushioned version of the pitman. (*American Machinist*)

pitman could be moved along the helve to increase or diminish the stroke. A subsequent patent, number 459,092 was granted to Vaughn on September 8, 1891. This patent included a modification of the pitman which readily enabled adjustment for varying thicknesses of work. The upper end of the pitman was threaded to pass through the block which connected it to the helve. The pitman could thus be rotated (by a wrench on its square upper end) to change the length of the rod between the crank and the helve. This also served to provide subtle adjustment of the stroke.

The second Vaughn patent was assigned to the Newark firm of F. J. Schmitt and Company, and it was in the hands of this company that the hammer was brought to market. A description of the hammer in *American Machinist* of October 8, 1891, mentions machines of 50 and 100 pounds, stating that "a number of these hammers are in operation." In addition to not requiring an expensive foundation, the hammers were said to "work very satisfactorily with a comparatively small power."[90] It is unlikely that the Vaughn power hammer was produced for more than a few years by F. J. Schmitt and Company. Apparently there were problems with the durability of the mechanism, and in 1896 Vaughn applied for another patent. The eventual rights to this patent were not assigned to Schmitt.

Vaughn's third power hammer patent was granted on June 28, 1898. Patent number 606,370 proposed a design intended to overcome "some defects which have developed and certain objections which have arisen in the practical operation of hammers made in accordance with my former patents."[91] Evidently the leather-strap linkage alone was insufficient to protect the drive mechanism from the jar of impact. Vaughn's new version included rubber or coil-spring cushions at the base of the pitman, where it connected to the crank. The fine adjustment of the stroke (by changing the length of the pitman) was still possible, but the adjustable element was relocated below the helve. The helve itself was redesigned (interestingly, much like the redesign of the Bradley Upright Strap Hammer's helve), with two steel semi-circular elements placed on edge, in lieu of the earlier single flat spring.

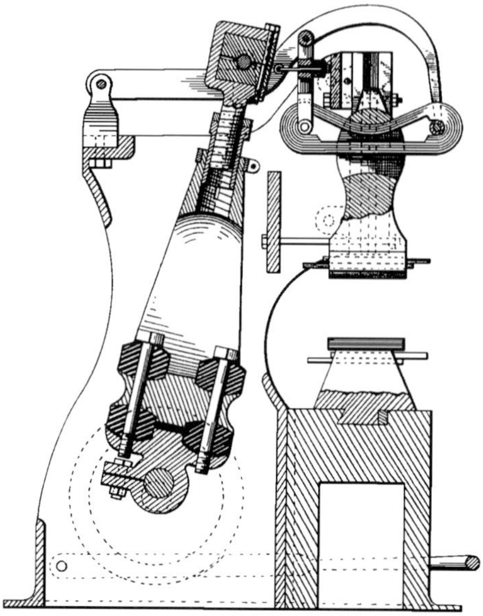

Charles Vaughn's third hammer patent was granted in 1898.

The pivot where the pitman connected to the helve was substantially reinforced and made adjustable to compensate for wear. Whether or not this hammer was actually built is not known. Vaughn's designs were clearly influential, and it is quite possible that there was a direct connection between Vaughn and the builders of several similar machines that soon appeared on the market.

* * *

R. E. Kidder, of Worchester, Massachusetts, manufactured a hammer which, in some respects, appears to be a redesigned Vaughn hammer. The Kidder Power Hammer was profiled in *The Journal of Railway Appliances* in June of 1897, at which time it was being built in six sizes from 25 to 200 pounds.[92] The key features of this machine are described in patent number 668,799 which was granted to Richard E. Kidder on February 26, 1901 (Kidder's application was

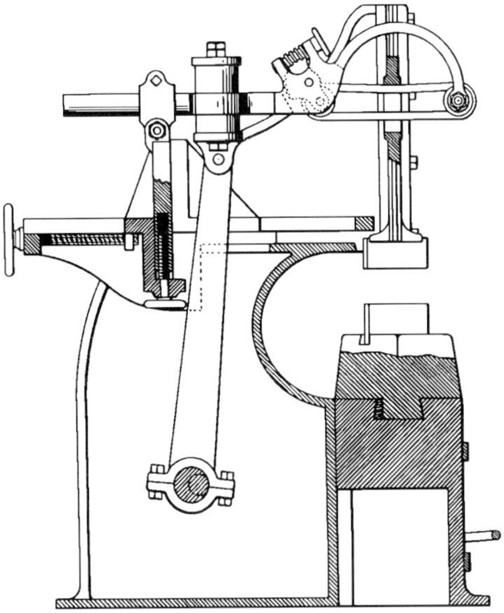

Richard Kidder's hammer was patented in 1901. Below, an advertisement from 1898. (*Machinery*)

filed in June of 1897, little more than a year after Vaughn filed the application for his improved design). Kidder was clearly aware of the shortcomings of Vaughn's earlier hammer, as is evident by the rubber cushions interposed at the connection of the pitman and the beam. However, instead of moving the connection point of the pitman to vary the hammer's stroke, Kidder chose to make the fulcrum adjustable. The fulcrum could be raised and lowered as well as moved forward and backward by turning handwheels. Furthermore, a mechanism was wisely provided for tensioning the leather strap that held the ram. Kidder's patent also described a brake, and a clever means by which the weight of the idler, or "binder," was removed from the foot treadle. The hammer was thus rendered less strenuous to operate, since the resistence of the belt was the only pressure to be overcome.

The firm of R. E. Kidder was established in 1869. Richard E. Kidder was a general machinist and patternmaker who manufactured a variety of machines. In addition to the power hammers, he built table saws and wood boring machines. After the turn of the century Kidder diversified into automobile repair. He died July 9, 1914 at the age of 72.[93] Manufacture of the power hammer may have continued briefly, for the company continued to be listed under "power hammers" in the 1915 edition of the *Thomas' Register of American Manufacturers*.[94] The fate of Kidder's Power Hammer after that date is not known.

* * *

Several years before the disappearance of Richard Kidder's hammer, another non-vertically configured hammer with a fulcrum at the end of the helve appeared on the market. The Central Machine Works Company, of Minneapolis, Minnesota, first appeared under "power hammers" in the 5th (1912) edition of the *Thomas' Register of American Manufacturers*.[95] An advertisement in *The Blacksmith and Wheelwright* from 1923 shows a far less sophisticated hammer than that built by Kidder. The fulcrum was

The Central Power Hammer ca. 1923. (*The Blacksmith & Wheelwright*)

fixed, and the ram was suspended from a conventional bow-spring and leather strap linkage.[96] As with the early Vaughn hammer, it was possible to adjust the stroke by sliding the connecting point of the pitman along the helve. The pitman was probably cushioned where it connected to the crank, as in Vaughn's 1898 design. A friction clutch took the place of the idler pulley used on both Vaughn's and Kidder's hammers. Surprisingly, the Central Machine Works Company was listed under "power hammers" in the *Thomas' Register of American Manufacturers* through 1990! However, by that date, the Central Power Hammer had been out of production for years. Apparently, no aspect of the design of this hammer was protected by patent.

The Central Machine Works Company was incorporated on February 1, 1900, with P. Monson as president, and J. Englund as secretary and treasurer.[97] By 1923, when it offered its hammers in *The Blacksmith and Wheelwright*, the company also advertised punch press work, screw machine products, castings, forgings, pattern-making, as well as the manufacture of special machinery.[98] John "Ken" Monson, grandson of founder Peter Monson, worked for the company until his retirement in 1985. He remembered, as a child of five, being terrified by the sounds of the hammer (an understandable reaction!) Interestingly, he remembered the earliest hammers as different from the one illustrated in 1923. Of the revised design he recalled,

> "The foot bar around the front of the machine operated a multi-disc cone clutch. The taper of the cone was such that it would not grab or lock-up. The operator could make the upper anvil just tap or he could force the upper anvil to raise up and by so doing the spring would exert an extra snap and make one hell of a blow."[99]

The company weathered the Great Depression and subsequent business cycles, becoming the Central Machine Corporation in 1968.[100] In 1984, Curtis Englund revived the old name at its original location (1234 Central Avenue, N.E., Minneapolis).[101] After seven years, the charter of the Central Machine Works Company was dissolved by the State of Minnesota.[102] Meanwhile, in 1984, the Central Machine Corporation became Tilt-Lock, Inc. In 1993, it merged to become TC / American Monorail, Inc.[103]

* * *

In February of 1898, an unusual advertisement appeared in *Machinery* magazine. B. A. Hjorth and Company of Stockholm, Sweden, offered to sell the patent for its Vulcanus Spring Power Hammer.[104] The U. S. patent in question, number 547,853 had

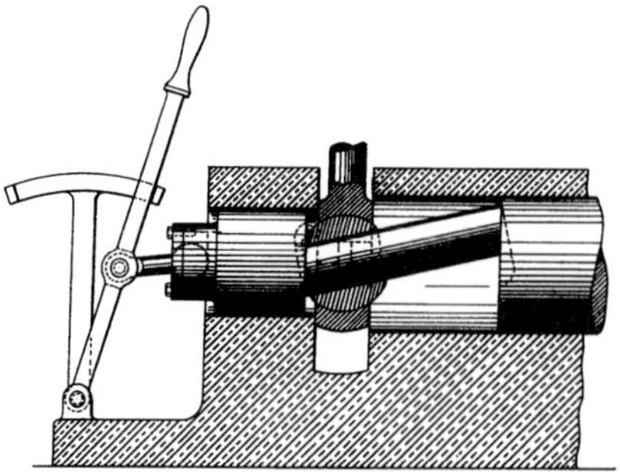

Johan Johansson's patented adjustable-stroke crank: as the piston moved inward, the wobble of the eccentric increased, increasing the stroke of the crank. This mechanism could be applied to other types of machinery; and, in fact, the patent makes no mention of hammers.

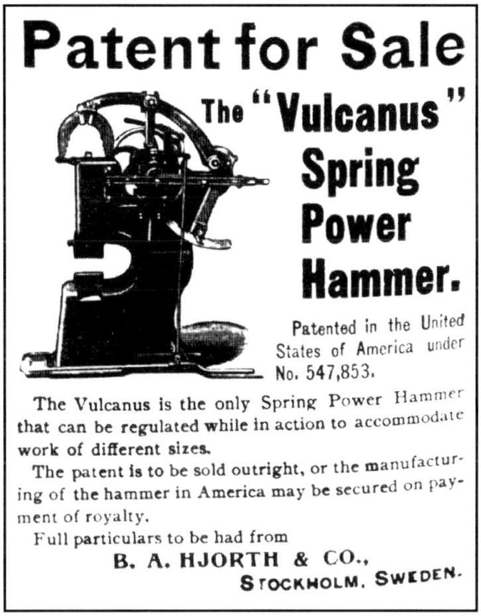

The Vulcanus hammer patent for sale. (*Machinery*)

been granted to Johan Petter Johansson of Enkjoping, Sweden, on October 15, 1895. It did not specifically refer to power hammers, but was instead a description of an adjustable-stroke crank. Johansson's device allowed subtle and precise regulation of a crank's stroke with a simple linear control lever. In addition to this undeniably convenient feature, the advertisement claimed that the hammer could be adjusted for varying thickness of work while the machine was in operation! Despite these merits, there was no further mention of the hammer, and it appears that the non-vertically-configured Vulcanus Spring Power Hammer was never manufactured in the United States.

* * *

The Kerrihard hammer was similar to the Central Power Hammer in several respects. It too had a stationary fulcrum located at the end of its beam, and the stroke of the hammer could readily be adjusted by moving the connection point of the pitman to the beam. Both machines had beams which were round in cross-section, and neither design was patented. Instead of relying upon the protection afforded by a patent, the Kerrihard Company of Red Oak, Iowa, simply produced a compact, high-quality hammer at an unbeatably low price. A full-page advertisement in *The Blacksmith and Wheelwright* from February, 1909, offered the Kerrihard hammer for just $60. More than 1,000 of these machines had been produced by that time, and the company pointed out that all of the initial expenses involved in creating patterns, jigs and special tooling needed to manufacture the hammer had long since been paid off. As a result, it was reasoned, a blacksmith could truly get his money's

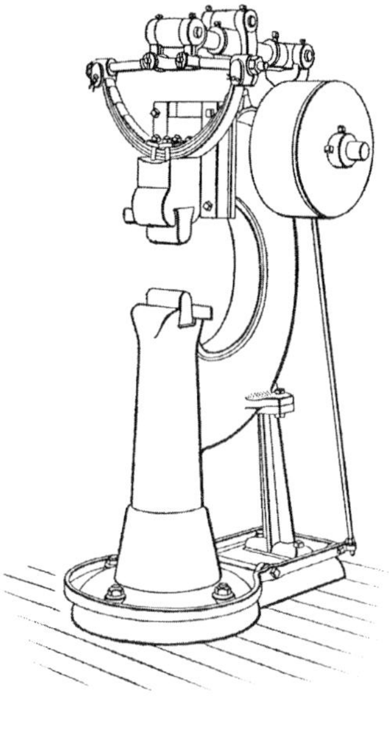

A line drawing of an early version of the Kerrihard hammer reveals an unusual upward orientation of the bow-spring linkage - also found on the earliest of Koch's Modern Power Hammers. (*International Library of Technology*, section copyright 1906) A redesigned Kerrihard hammer cost a mere $60 in 1909. (*Kerrihard Company trade literature*)

worth. The smith was also reminded of the uncertainties involved in buying a "theoretically correct" hammer which was still in the experimental stage. The 1909 Improved Power Hammer represented, it was claimed, the culmination of seven years of development. The continuing popularity of the hammer was supported by the claim that they were being produced in Red Oak at an average rate of five per day.[105] The Kerrihard hammer was indeed highly successful and outlived the majority of its competitors.

The stroke of the Kerrihard hammer was easily adjusted by loosening a bolt and shifting the pitman along the round steel beam. With the pitman all the way forward, the Kerrihard could handle even thick, stubborn plow shares. (*Kerrihard Company catalogue*)

The 30-pound, 1909 model Kerrihard hammer was among the smallest of the many crank-actuated, guided-ram hammers manufactured in the United States. It stood 55 inches high and weighed only 700 pounds overall. The hammer used an extremely compact pivoting-beam arrangement, with motion provided through a yoke attached to a crank shaft. The yoke could be moved along the round steel beam to adjust the length of the stroke. A dependable friction clutch was mounted to the crank shaft, eliminating the need for a countershaft. The linkage to the ram employed a single coil spring, much like Dupont's design. The compact configuration of the Kerrihard hammer made it ideally suited to tire-welding. Even tires of relatively small diameter could encircle the hammer mechanism. In addition to the 30-pound model, a No.2 Improved Power Hammer was also offered. It stood five inches taller, weighed 1,250 pounds overall, boasted a 75-pound ram, and sold for $115.[106] Trade literature from the manufacturers boasted that it sold "the hammer that pounds out the profits."

The Kerrihard Company was founded in 1890 as a general plumbing business.[107] It flourished under the direction of E. W. Kerrihard, the son of the founder. An advertisement in the Christmas issue of the *Red Oak Independent* solicited all types of plumbing work and offered to make one's house "a perpetual summer, by warming it with a hot water or steam heating apparatus."[108] The company added a foundry and began to produce its own pumps and a variety of other products. Kerrihard also advertised its line of galvanized sheet-metal products extensively, including watering tanks, grain bins and hog troughs. At the turn of the century, the company employed from ten to fifteen men. Perceiving an opportunity in the difference between rail costs for raw, pig iron and finished goods, the company began to produce soil pipe as well.[109] A new foundry was installed in October of 1901,[110] and hammer production began within the next year. By the Spring of 1907, business was strong enough to warrant another expansion. A new building and a cupola of eight tons melting capacity were added. Two dozen men were employed at this time, and Kerrihard began to sell a machine for beveling cardboard and mattboard. E. W. Kerrihard designed this machine initially for the best-known of Red Oak's industries: the Thomas D. Murphy Art Calendar Company. The advertising calendars made by this firm were well-known throughout the United States during these decades.[111] The Kerrihard Company further diversified its line with the addition of emery grinders, "front-pul" shears, disc rollers and saw arbors.[112] The power hammer was an important Kerrihard product for many years, and remained largely unchanged until the tough market during the Depression prompted a modernization.

After a seven-year absence from the pages of *The Blacksmith and Wheelwright*, advertisements for Kerrihard's new, "C-30"

The Kerrihard C-30. (*The Blacksmith & Wheelwright*)

hammer appeared in May of 1930. The redesigned Kerrihard hammer reflected changing aesthetics in machine design, with the base, column and anvil cast as one unit. To quote a Kerrihard advertisement: "There is maintained compact rigid construction, which is so rapidly superceeding the sectional methods of all machine tool design."[113] The coil spring in the earlier linkage was replaced with rubber cushions in a linkage reminiscent of that employed on the Bradley Compact hammer. In another change, the dies were cleverly fashioned from round tool-steel stock, making them easily replaced, modified, and rotated in both the anvil and ram. Despite these changes, the hammer remained small, compact, and highly affordable. The C-30 sold for $105; and with speeds of 300 to 400 blows per minute, boasted that it could hit faster than any other hammer made.[114] The new hammers were evidently successful, for the Kerrihard Company continued to be listed under "power hammers" in the *Thomas' Register of American Manufacturers* through 1945. By that time, though, the company was devoting the majority of its energy to the manufacture of one of its earliest products: heating furnaces. Reflecting that emphasis, the company changed its name to the "Ko-Z-Aire Manufacturing Company" on April 16, 1947. During the next few years there were several changes of ownership. In 1953, as a subsidiary of Anchor Post Corp., the plant was building 35 different models of furnaces, with an annual output of 10,000 units.[115] By then, the Kerrihard hammer was all but forgotten.

* * *

The Sterling Hammer appeared in the "New Tools and Machine Shop Appliances" column of *American Machinist* in July of 1905. In a single 40-pound size, it was offered specifically "for

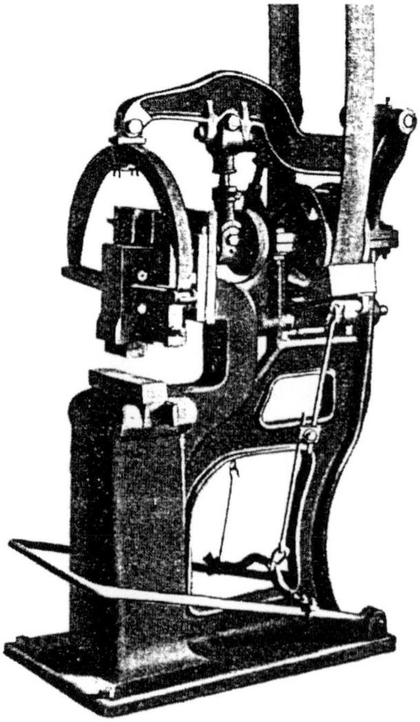

The Sterling Hammer. (*American Machinist*)

carriage makers and manufacturers of small iron and steel goods."[116] It stood four feet nine inches tall and weighed 1,700 pounds overall (shipping weight). The crankshaft was "placed low to avoid vibration." While it resembled the Kerrihard, the connection point of the pitman to the beam was not adjustable. The frame was playfully curvilinear and reminiscent of the Palmer helve hammer.

The Sterling Hammer was built by the Sterling Machine Works of Sterling, Illinois. Aside from assurances that the hammer was designed by a practical blacksmith, there does not appear to be a patent associated with this machine, so the identity of the inventor is uncertain. The company first appeared in the 1901 edition of the *Twentieth Century Directory of the Cities of Sterling and Rock Falls, Illinois*. The principles were John and Peter Frantz.[117] In the 1904 directory, an advertisement for the company lists Peter Frantz as the proprietor. While the Sterling Machine Works would take on general repairing of all kinds, it was noted that the firm's specialty was the invention and building of wire-bending machinery.[118] By 1906, after the appearance of the power hammer, G. Cassens is listed as the proprietor.[119] Several years later G. Cassons is described as treasurer, with E. M. Cobb president, H. M. Cosey vice president, and Earl Kelly secretary. Though Peter Frantz left the company, he did not leave Sterling, for he later founded the Frantz Manufacturing Company. In addition to the hammer, an interesting product of the Sterling Machine Works was automobiles. In June of 1909, the *Sterling Gazette* reported that the first two automobiles from the Machine Works had taken to the streets. Built along the line of a common "spring wagon" these horseless carriages were intended for light hauling and pleasure use.[120] At this time, the hammer was still in production. An advertisement for the Sterling Hammer appeared in June of 1910 in *The Blacksmith and Wheelwright*. It was boasted that it had "fewer parts than any other hammer of equal capacity."[121] The company did appear in the fifth (1912) edition of the *Thomas' Register of American Manufacturers*, but since it was not included in the city directory of that date, the *Thomas' Register* listing may have been posthumous.[122]

* * *

The hammer designed by Austin Bragg, of Waterville, Maine, was claimed to offer "the nearest approach to the smith's arm attainable by mechanical means."[123] This was accomplished with a compound leaf spring helve which was reminiscent of Joseph Palmer's hammer; and which may have, in turn, influenced Henry Alinder. The design also included a means for adjusting the hammer's stroke with the machine either idle or in operation. An eccentric located beneath the fulcrum could be turned to raise or lower the fulcrum to accomodate work of varying thickness. The fulcrum was adjusted by turning a conveniently-located handwheel which was connected to the eccentric by a long shaft, sprockets and chain, and a worm gear. Bragg claimed that it was thereby possible to regulate the stroke of the ram "to a hair's breadth."[124]

With a seven-year hiatus, Bragg's Hammer appeared twice as a new product. (*American Machinist*)

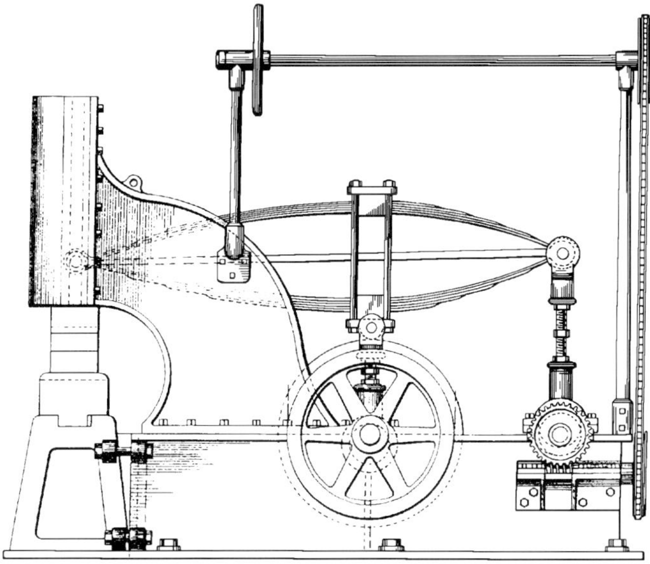

Bragg's hammer was patented in 1906. The fulcrum of the beam could be readily adjusted with the hammer either idle or in operation.

Bragg was granted patent number 810,905 on January 30, 1906. By that date, the hammer had already been in production for several years. The hammer was profiled by *American Machinist* in September of 1903, at which time it was being built in one size by Austin Bragg himself. The machine had a ram weighing 100 pounds, weighed 4,500 pounds overall, and ran at a maximum speed of 275 blows per minute. The article notes that the ram traveled in bronze guides which were constructed in one piece, and which also provided a cover for the ram. In addition to the adjustment of the stroke previously described, it was possible to adjust the tension on the compound spring helve to suit various types of work.[125]

Seven years later, the Bragg design reappeared in the pages of the *American Machinist* as a New Power Hammer. The machine was described as a recent product of the New Metal Tool Steel Company, of Standish, Maine. It was recommended particularly to those having interchangable parts to forge, being capable of forging to within very close limits.[126] Neither Bragg nor the New Metal Tool Steel Company were ever listed in the *Thomas' Register of American Manufacturers*, and it is unlikely that these hammers were built in great numbers.

* * *

The last non-vertically-configured guided-ram hammer to be considered here is among the most interesting, mechanically. The hammer developed by James Sweeney and Robert Laird of St. Johnsbury, Vermont, was intended to replicate the "elastic" quality of a hand hammer blow. It employed a distinctive mechanism which reduced vibration and shock. It also permitted adjustment of both the stroke and the capacity of the hammer while it was in

The Laird & Sweeney Power Hammer in 1893. (*The Iron Age*)

operation. In a highly compact and effective design, the drive shaft and the mechanism for adjusting the hammer turned upon the same center. This permitted counter-balancing which rendered the hammer extremely efficient, easily adjusted and smooth in operation. So little power was required to operate hammers of this design that the inventors anticipated that they could also be worked by either hand or foot power!

James Bernard Sweeney and Robert Wallace Laird were granted patent number 477,159 for this design on June 14, 1892.* The hammer ram was oscillated by a centrally-pivoting beam. The connection of the beam to the ram was also highly unusual: extending from the front end of the beam was a strip of spring steel;

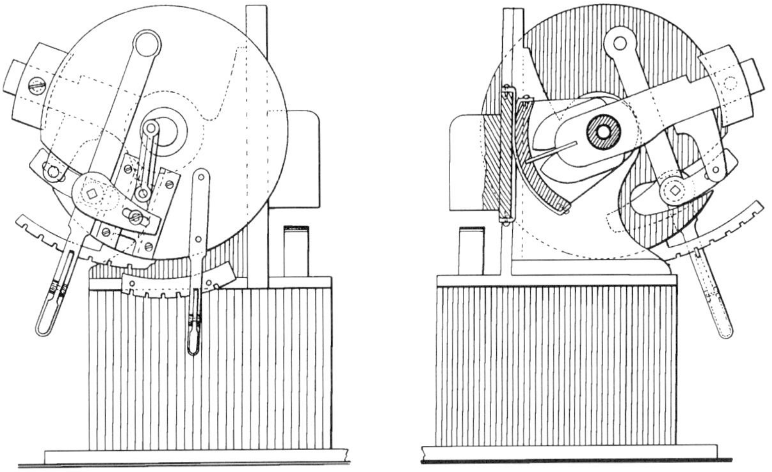

James Sweeney and Robert Laird's first patent of 1892: a full disc was used to illustrate the stroke-adjusting mechanism, which shared its center with the drive shaft.

to this was attached a curved plate; straps were attached to each end of this curved plate, which crossed each other to be attached to the upper and lower ends of the ram. This arrangement provided a positive, yet flexible, movement of the ram in both directions. The

* Since Sweeney is listed first in the patents, and Laird has priority in the company name, their respective roles can be deduced.

novel actuating and adjusting mechanism was illustrated in this first patent with a full disc, though unnecessary portions of that disc were eliminated in the development of the actual hammers. At the center of the disc ran the drive shaft with its crank. The disc bore a slide which converted the rotary motion of the crank to linear. The disc could be rotated slightly and locked in place (with a hand lever and rack of detents) thereby raising or lowering the ram and adjusting the hammer for various thicknesses of work. A second hand-lever and rack varied the linkage from the linear slide to the beam. This adjusted the stroke, from the lightest of blows to the heaviest (heavy blows being delivered with the lever in its furthest backward setting). With an assistant operating the hammer's levers, the smith could commence an operation with heavy blows, and steadily diminish the height of the ram and its stroke as the work progressed. The inventors did not limit the application of their invention to forging, and specifically mentioned the potential use of their balanced hammer as the drive for a (rock) drill.[127]

The Laird and Sweeney Power Hammer, as built by the Laird and Sweeney Manufacturing Company, was described in *The Iron Age* in February of 1893. Though unnecessary portions of the

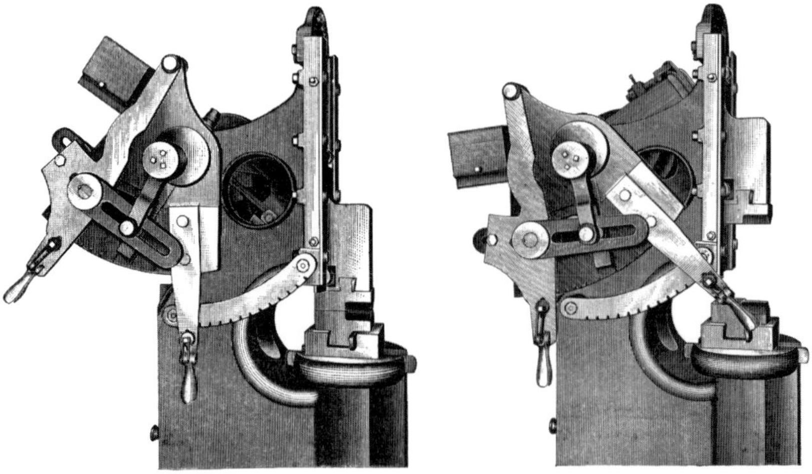

The Laird & Sweeney Power Hammer could be adjusted for varying thicknesses of metal while the machine was running. (*The Iron Age*)

disc were cut away, the mechanism operated substantially as described above. The hammer illustrated in this article (of unspecified weight) offered a range of stroke from two to eight inches with an opening between the dies of 16 inches. The heavy anvil was cast separately from the frame, and the overall machine weighed 1,500 pounds. Additionally, "the forged steel shaft has a bearing of 24 inches through the hub. These hammers are now built in sizes from 25 to 500 pounds weight of head."[128] This was a sophisticated machine with many precision parts, and was probably expensive to manufacture and purchase. With that in mind, the inventors' subsequent "improvements" are understandable (though perhaps a bit disappointing, given the elegance of the original design).

On February 13, 1894, Sweeney and Laird were granted patent number 514,787 for a simplified version of the above mechanism. The height of the ram was fixed with a handwheel, which positioned a rocking link along a curved track. The stroke was adjusted by moving a connecting link (which was attached to the rocking link) nearer-to or farther-from the beam's pivot. This

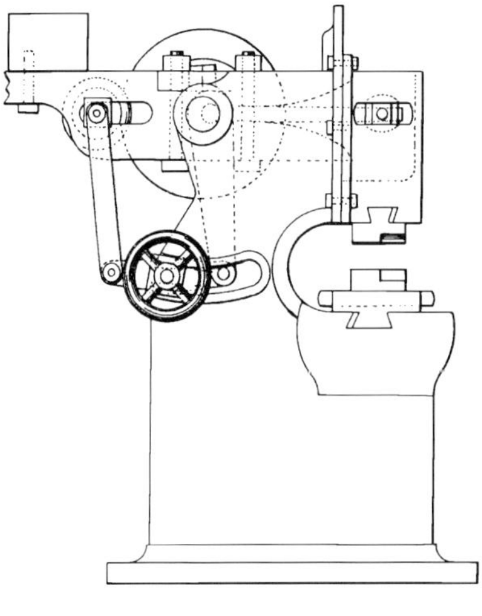

James Sweeney & Robert Laird's patent of February, 1894, described a simplification of the original design. The stroke could no longer be adjusted while the hammer was running.

was a more conventional method of varying a forging hammer's stroke, and was probably far more economical to build. While the capacity of the hammer could still (probably) be adjusted while the hammer was in operation, it would have been impossible to similarly adjust the stroke.

The process of simplification continued in their third patent, number 529,634 granted November 20, 1894. In this design, the stroke was varied by moving the connection of the pitman to the beam (near its end), by meams of a sliding "clip" held

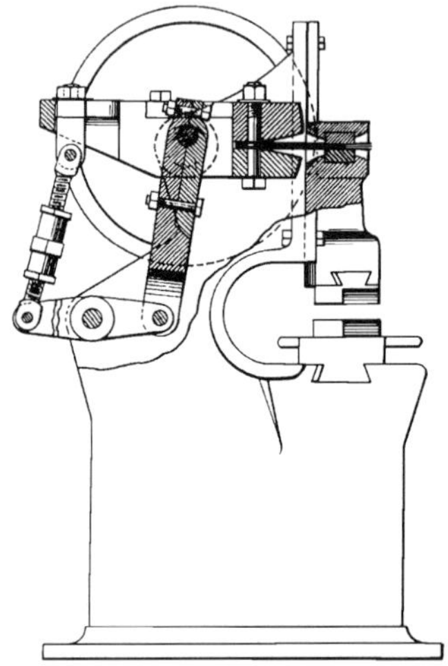

James Sweeney & Robert Laird's third patent described a hammer which could not be adjusted at all without stopping the machine.

with a bolt. The capacity of the hammer was regulated by an adjustable pitman. Neither of these adjustments could be made with the hammer in operation. There was one additional facet to this patent which is most unusual: the inventors proposed a subtle regulation of the hammer's elastic stroke by varying the number of spring plates which were used to connect the ram to the beam (directly, without the original straps). It is not known if hammers with these modifications were ever manufactured.

The fate of the Laird and Sweeney Manufacturing Company, and its remarkable hammer, is not known. The company was never listed in the *Thomas' Register of American Manufacturers* (nor was it mentioned among industrial enterprises (past or present) in Edward T. Fairbank's *The Town of St. Johnsbury VT* (published in 1912). The family name of Laird does appear in St. Johnsbury, in connection with the quarrying of granite, but there is no mention of Robert Wallace Laird.[129] It is possible that the Laird and Sweeney hammers were produced in the local shop of O. V. Hooker & Son, the same works which built the first Dupont hammers several years earlier.

* * *

> The Smith his iron measures,
> Hammered to the anvil's chime.
> - *Longfellow*
>
> And his **Beaudry Hammer** treasures,
> Since it saves him work and time.
> - *Some-other-fellow*

Beaudry Company catalogue, ca. 1912

A Little Giant Power Hammer driven by a small gasoline engine, ca. 1903. (*The Blacksmith & Wheelwright*)

CHAPTER 6

"Hooker to Your Ingin"

As was said in Chapter 1, crank-actuated open-die power forging hammers have again assumed an important place in the shop of the modern blacksmith. Now, at the close of the twentieth century, these machines have been out of production for decades, and yet an appreciation of their versatility and potential as tools for artistic expression is steadily on the rise. Crank-actuated hammers still spare the blacksmith the cost of expensive human labor and perform the hardest pounding, but they can also facilitate the sculptor's exploration of form and texture. The speed and ease of operation of such hammers gives the artist-blacksmith increased access to the process of forging and encourages exploration. *This remarkable new role for power forging hammers prompted the present investigation into their origins; but the breadth and scope of those origins proved to be even more remarkable.*

Crank-actuated power forging hammers appeared as a distinctly American contribution to the history of forging machinery. Over the course of roughly a century, hundreds of American inventors applied their genius to improving what was fundamentally a simple machine. The seriousness of their intentions is evident in the U. S. patent record (and few hammer designs were placed on the market without patent). Indeed, a survey of the patents dedicated to this single machine provides a wonderful testament to human ingenuity. Dozens of American companies across the country applied their resources to the manufacture of these hammers. The story of each of these companies has been one of flesh and blood as well as foundry sand and metal shavings. While it is not popular to view such mundane activity as heroic, each of these firms, and the hammers which they built, were important to many lives. The owners, employees and the growing communities in which the companies were situated all depended upon their prosperity. The history of this industry also reflects important changes that occurred in America as it rose to industrial prominence. Early crank-actuated hammers were primarily a manufacturer's tool. With the new availability of small gasoline engines at the turn of the century, power hammers came within reach of even the rural, repair-oriented blacksmith. The profusion of hammers which appeared in the early decades of the twentieth century further suggest that the blacksmiths' profession was at a peak, not on a long path of steady decline. However, when that decline inevitably came, it was swift. The obscurity to which crank-actuated power forging hammers descended (and the extent to which they have been ignored by industrial historians) must be attributed in part to the speed and thoroughness with which the blacksmiths' trade became all but superfluous in America. The triumph of the automobile over horse-drawn transportation was undeniably a principle factor in the decline of the general blacksmith's trade (farriers continue to flourish, of course, in their specialized niche). That process was already well under way by the end of the First World War, but there were many other developments which came together at that point in time. None of these, however, struck at the heart of the trade and its craft, as deeply as the widespread adoption of alternative welding processes.

Traditionally speaking, one of the most useful and virtually magical capabilities of the blacksmith was his ability to weld at the forge. Until the beginning of the twentieth century, manufacturers and engineers were greatly limited in the ways in which ferrous materials could be joined. Parts could be held together mechanically, by rivets or screws, or they could be welded together at the forge. In its most dramatic application, forge welding was also used for building-up large masses of iron for articles such as anchors, engine shafts and heavy ordnance. On a smaller scale, forge welding was widely used in blacksmiths' shops as a repair technique. Replacement tires for wagon wheels were welded into continuous bands at even the most humble rural forge. Worn axle spindles were cut off and new stub axles welded in their place. Plowshares worn beyond a simple sharpening were given new life by forge-welding additional metal onto the share. These were, of course, the three operations most frequently mentioned in the promotion of power hammers; but further applications of the forge welding process were practically unlimited. Most blacksmiths were adept at the fundamentals of forge welding, but the process was demanding and the results were less than certain. Difficulty increased in the welding of dissimilar metals (such as steel to wrought iron) and in welding sections of unequal mass. Many metals could not be welded at the forge at all due to the extended exposure of the material to oxygen at an elevated temperature. By the end of the nineteenth century a number of promising alternatives to forge welding had been proposed, technologies which both simplified the welding process and extended the range of materials which could be welded.

One of the earliest alternative welding processes to be developed was electrical resistance welding. In electrical resistance welding a current is passed across a joint, and the resistance at that point concentrates heat and produces the weld. Elihu Thompson first conceived of the process in 1877. Following experiments in 1883-1885, he obtained his first patent in 1886. While the technology was almost fully developed from the start, it was not rapidly adopted.[1] Among the earliest resistance welding machines were those designed to "butt-weld" continuous steel hoops, such as tires for vehicle wheels. This new technology was well-suited to factory mass-production; but forge welding, by hand or power hammer, remained more practical in repair circumstances for several decades.

Electric-arc welding was first employed commercially in 1887. Though successfully used to repair metal tanks and similar products, its initial application was very limited. At that time it was impossible to weld at voltages over 150 amperes, and few materials lent themselves to welding by this process. By the early 1890's, an English firm had installed a carbon-arc welding outfit of 550 amperes with three dynamos. Though improved, the equipment was extremely bulky: 1,800 storage battery cells were required to handle the peak load! In 1910, a number of commercial arc-welding plants were established, but the process was still limited by the use of rods of very low carbon content. Rods of various compositions (to suit the metallurgical requirements of diverse applications) appeared on the market in 1915; and the arc welding process became widely accepted soon after the First World War.[2] Scientific research into the composition of welding rods led to the development of so-called hard-facing alloys. Applied with an electric arc to surfaces such as the edges of plowshares , these alloys greatly extended the life of the tool. However, hard-faced tools could not be repaired or modified at the forge.

Oxy-acetylene gas welding became feasible just after the beginning of the twentieth century with the development of a suitable torch. Acetylene had been discovered in 1836, but its manufacture and use did not occur until the 1890s. In 1892, calcium carbonate was created by accident in an electric furnace at the Willson Aluminum Company in North Carolina. It being a more naive and industrially-rapacious era, this unwanted substance was thrown into the river. It was subsequently discovered that the evolving gas was flamable. At first, acetylene gas was developed for illumination purposes, but it was soon recognized that the oxy-acetylene flame burned hotter than any other known gas or solid fuel, and was thus well-suited to welding and cutting.[3] The first industrial application of the process occured in 1903. Like arc welding, though, its use was not widespread until just after World War I. Early oxy-acetylene welding outfits included their own acetylene generators; but, within a decade, a method had been devised for dissolving the acetylene in acetone. The dissolved acetylene could be easily transported and dispensed from cylinders. By the 1920s, the Prest-O-Lite Company, which supplied acetylene in this form, claimed that "the oxy-acetylene welder's hand-truck is the ambulance of the modern mechanic's shop."[4] The process not only replaced the blacksmith in one of his most respected roles, but greatly expanded the scope of what could be welded. The additional ability to cut with a heavily oxidizing flame proved extremely useful. Cutting torches helped to process the scrap metal which was collected feverishly during both World Wars to fuel wartime industrial production (this included, of course, the destruction of many of the older crank-actuated power forging hammers mentioned in the preceeding chapters). While tradition-oriented modern blacksmiths may feel more than slightly ambivalent toward electric and gas welding processes, the majority of blacksmiths today have embraced these resources.

* * *

Throughout the centuries-old story of the blacksmith's craft there have been artists, but only recently has the very survival of the craft depended upon an appreciation for the expressive potential intrinsic to its processes. Crank-actuated power forging hammers pre-date that transformation, but in their inherent versatility they have once again become valuable tools at the forge. Blacksmithing is no longer a conventional profession, one in which a son might well be expected to follow in his father's footsteps. Many of today's blacksmiths have been introduced to the craft through academic programs, often emphasizing artistic expression. Others have entered the craft through a love of its history and traditional processes. Traditionally-minded smiths have been indelibly influenced by the expressive work of their peers, just as the artist-blacksmith has profited by exposure to the highest standards of traditional craftsmanship. The story of the twenty-first century blacksmith, and his or her accomplishments using technologies new and old, promises to be rich, indeed! Knowledge of the history of the blacksmith's processes and tools (including crank-actuated power hammers) will continue to enhance the pleasure and potential to be found at the forge.

Notes

Chapter 1 - *Opening Remarks*

[1] Manning, Maxwell & Moore Co., *Illustrated Catalogue of Railway & Machinists' Tools* (New York, Jan., 1894), p. 286,288. (Stanford)

[2] Kerrihard ad., *The Blacksmith & Wheelwright*,Vol. 59, #2, Feb., 1909, p. 575.

[3] Brad Kava, "Last of his kind - He forges farm tools the old fashioned way," *San Jose Mercury News*, Jan. 27, 1896, p. B1 & B3.

[4] Williams, White & Co., *The First Hundred Years* (Moline, Ill., 1954), p. 36.

[5] C. Chomienne, "Notes on Steam Hammers," *The Railroad & Engineering Journal*, Nov., 1888, pp. 500-502.

[6] "The Bethlehem 125-ton Hammer," *The Engineer*, Vol. 76, Nov. 24, 1893, pp. 480-481.

[7] Victor S. Clark, *History of Manufacturers in the United States*, Vol. 2. (New York, McGraw Hill, 1929) p. 80.

[8] The inclusion of Olivers (the smallest and least sophisticated of open-die power hammers) in the same general category with direct-acting steam-powered hammers (the largest and most dramatic of hammers) warrants further discussion. It so happens that all foot-powered hammers of which the author is aware (in both patent and actual form) employ a simple, "direct" means of communicating motion to the hammer. It is entirely possible to envision a foot-powered hammer which could use a trip or crank actuation (perhaps via a flywheel). In such an instance, the hammer would belong to the appropriate category, not "direct." The inclusion of Olivers in the "direct" category is thus not a function of their source of power.

[9] Hurt, Douglas, *American Farm Tools from Hand-power to Steam-power* (Manhattan, Kansas, Sunflower University Press, 1982), p. 9.

[10] John W. Oliver, *History of American Technology* (New York, Ronald Press, 1956), p. 224.

[11] John J. Thomas, *Farm Implements & Farm Machinery* (New York, Orange Judd, 1869), p. 120.

[12] Nicholson File Company, *File Filosophy* (Providence, R.I., 1943), p. 46.

[13] Crucible Steel Co. of America, catalogue (New York, 1925), pp. 120-121,150.

[14] Reichelt, Richard, et al., *Heartland Blacksmiths* (Carbondale, Ill., Southern Illinois University, 1988), pp. 28-29.

[15] Cray Brothers, *Net Price Catalogue 1909* (Cleveland, March, 1909), p. 175. (author's library)

[16] "German vs. American Trip Hammers," *American Machinist*, Vol. 2, Feb., 1879, pp. 6-7.

[17] Robert L. Merriam, et al., *The History of the John Russell Cutlery Company* (Greenfield, Mass., Bette Press, 1976), p. 30.

[18] "Trip hammers. To what extent are they useful in blacksmithing?" prize topic, *The Blacksmith & Wheelwright*, Vol. 49, #1, Jan., 1904, p. 21.

* * *

Chapter 2 - *Atmospheric Hammers*

[1] Overman, Frederick, *The Manufacture of Iron in All its Various Branches* (Philadelphia, Henry C. Baird, 1851), p. 337.

[2] Cox, Tench, A *Statement of the Arts and Manufactures of the United States of America for the Year 1810* (Philadelphia, U. S. Treasury Department), p. xxxi.

[3] Roe, Joseph Wickam, *English and American Tool Builders* (New Haven, Yale University Press, 1916), p. 143.

[4] U.S. Patent, Nov. 25, 1851.

[5] Spaulding, B. F., "Ramrods and Spindles at the Trip-hammer," Practical Details of Blacksmithing, *American Machinist*, Vol. 11, No. 13, March 31, 1888, p. 4.

[6] Smiles, Samuel, ed., *James Nasmyth - an Autobiography* (New York, Harper and Brothers, 1883), p 253.

[7] Overman, p. 337.

[8] "Improved Atmospheric Hammer," *Scientific American*, Vol. 1, No. 20, Nov. 12, 1859, p. 1.

[9] "Hughes' Atmospheric Trip Hammer." *Scientific American*, Vol. X, No. 9, Nov. 11, 1854, p.1,

[10] "Improved Atmospheric Hammer."

[11] Ibid.

[12] "Atmospheric Hammer," *The Engineer*, Vol. 8, 1859, p. 412.

[13] "Waterhouse's Fallhammer mit Luft-Reitel ober Luft Breller," *Dr. Dingler's Polytechnischem Journal* (Berlin), Vol. CLI, 1859, p.253-254 and plate IV.

[14] "Dawes and Carr's Atmospheric Hammers," *The Engineer*, Vol. 9, March 9, 1860, p. 156.

[15] Rosenberg, Nathan, ed., *The American System of Manufactures - Report of the Committee on the Machinery of the United States* (Edinburgh, Edinburgh University Press, 1969), p. 104.

[16] *Commerce, Manufactures & Resources of New Haven, Conn.* (New Haven, National Publishing Co., 1882), p. 25.

[17] *New Haven City Directory*, 1857.

[18] U. S. Patent no. 27,925 (granted April 17, 1860), p. 1, lines 32-50.

[19] ibid., p. 2, lines 4-14.

[20] According to a genealogy (no author) of the descendants of Samuel Hotchkiss (in the collection of the Whitney Library of the New Haven Colony Historical Society), Caleb Hotchkiss, who was the great, great, great, great grandfather of Bennet, married Elizabeth, daughter of Henry Peck. A long stretch perhaps...

[21] ibid.; It is quite possible that Bennet Hotchkiss made further contributions to the field of drop hammer design. A board drop included among the machinery displayed at the Centennial Exposition was described as "The Hotchkiss and N. C. Stiles' Friction Roll Drop: Stiles and Parker Press Company." (*The Masterpieces of the Centennial International Exposition*, Vol. III, p. 371).

[22] U. S. Patent re-issue no. 7,475 (filed Aug. 5, 1876, granted Jan. 23,1877), p. 1.

[23] "Hotchkiss' Atmospheric Trip Hammer," *The Engineer*, Vol. 15, May 1, 1863, p. 247.

[24] "Hotchkiss Atmospheric Forge Hammer," *Scientific American*, Vol. 11, No. 14, Oct. 1, 1864, p. 1.

[25] "Improved Atmospheric Hammer," *Scientific American*, Vol. 13, Sept. 16,1865, p. 178.

[26] "Dead-stroke Power Hammer," *Scientific American*, Vol. 15, 1866, p. 134.

[27] U. S. Patent no. 51,267 (granted Nov. 28, 1865), p. 1.

[28] "Dead Stroke Power Hammer."

[29] "Improved Forge Hammer," *The Scientific American*, Vol. 14, 1866, p. 258.

[30] *The Journal of Railway Appliances and Railway Price Current* (New York), Vol. XII, Number 12, Dec., 1892, p. 54.

[31] *Thomas Register of American Manufacturers*, 4th edition, August, 1909, p. 957.

[32] Pratt & Whitney Company, *Illustrated Catalogue and Price List*, Hartford, Connecticut, July, 1881, p. 4. (Research Center Library, Henry Ford Museum and Greenfield Village).

[33] Scott, John, *Genius Rewarded, or the Story of the Sewing Machine*, 1880.

[34] Hutton, F. R., *Report on Machine-tools and Wood-working Machinery*, Tenth U.S. Census (Washington, Department of the Interior - Census Office, 1885), p. 10.

*   *   *

Chapter 3 - *Crank-actuated Helve Hammers*

1 *The Scientific American*, Vol. V, No. 1 (New Series), July 6, 1861, p. 1.

2 U. S. Patent no. 30,343 (granted Oct. 9, 1860), p. 2, lines 40-44.

3 American Society of Mechanical Engineers, *Mechanical Engineers in America Born Prior To 1861 - A Biographical Dictionary* (A.S.M.E., New York, 1980), p. 115.

4 "Cornelius H. De Lamater," *American Machinist*, Vol. 12, No.7, Feb. 14, 1889, p.7.

5 "Cossit's Corners Looked to a Lad Like Land of Promise," *Syracuse Post-Standard*, May 7, 1922; and, "Death Summons C.C. Bradley, Sr.," *Syracuse Post-Standard*, Jan. 29, 1916.

6 "Bradley's Cushioned Hammer," *Scientific American*, Vol. 28, March 22, 1873, p. 182.

7 "Mechanical Refinements No. XVII," *Engineering* (London), August 16, 1872, p. 110. Penned by "An English Engineer in America."

8 Bradley Mfg. Co. ad., *The Iron Age*, Feb. 20, 1873.

9 John W. Oliver, *History of American Technology* (Ronald Press Co., New York, 1956), p. 475.

10 *Engineering*, Vol. 22 , Oct. 20, 1876, p. 334.

11 *American Machinist*, Vol. 4, June 4, 1881, p. 9.

12 Manning, Maxwell & Moore, *Illustrated Catalogue of Railway and Machinists' Tools & Supplies*, New York, Sept., 1884, p. 296. (author's library)

13 ibid.

14 ibid. p. 296A.

15 Richard Dudgeon, Bill of Sale dated May 20, 1881, for a 6" steam hammer sold to the New York City Fire Department. Hold-down bolts and timbers cost an additional $30. (author's library)

16 Kerrihard Co. ad., *The Blacksmith & Wheelwright*,Vol. 59, No. 2., Feb., 1909, p.575.

17 C.C. Bradley & Son, *The Bradley Rubber Cushioned Helve Hammer* (brochure), 1911, p. 4. (Chambersburg Engineering Company archives)

18 *American Machinist*, Vol. 1, Aug. 2, 1879, p. 10.

19 ibid., p. 8.

20 ibid., Vol. 4, Dec 3, 1881, p. 9.

21 ibid., Vol. 5, July 15, 1882.

22 ibid.,Vol. 10, April 16, 1887, p. 9.

23 *Syracuse N. Y., Illustrated*, 1888, p. 43.

[24] Bradley & Co. ad., *American Machinist*, May 9, 1889, Vol. 12, No. 19, p. 10.

[25] "Bradley's Hard Coal Heating Forges," *The Blacksmith & Wheelwright*, Vol. 4, No. 3, Sept., 1881, p. 238-239.

[26] "Death of C. C. Bradley," *The Blacksmith & Wheelwright*, March, 1916, p. 97.

[27] ibid.

[28] *Machinery*, March, 1918, p. 179.

[29] ibid., Feb., 1919, p. 339.

[30] C.C. Bradley & Son, Inc., *New Bradley Developments, No. 5*, ca. 1948.

[31] ibid., No. 2, p. 2.

[32] "Shots Fatal to Former City Officer," *Syracuse Post-Standard* (?), March 9, 1932.

[33] "Cortland Machinery Firm Plays Important Role in Nation's Defense Program," *Syracuse Post-Standard*, Jan. 26, 1941.

[34] "Edlunds Turns Out New $35,000 Machine that Thinks for Itself In Drilling, Tapping, Boring," *Cortland Standard*, July, 15, 1960

[35] Joyce Smith (of Cortland Machine & Tool Co., Inc.), letter of May 12, 1994. The Cortland Machine & Tool Co., Inc.is located at 60 Grant St. and may be reached at P.O. Box 27, Cortland, New York, 13045.

[36] "The Long & Allstatter Company - Power Punches and Shears," *The Republican-News* (a special "Greater Hamilton Edition," n.d., Winter, 1909), p. 44.

[37] Long & Allstatter Company, *Illustrated Catalogue of Power Punches, Shears and Hammers*, Hanmilton, Ohio, n.d.,ca. 1873, pp. 22-23. (author's library)

[38] Double hammers were built in practically every hammer type (foot-powered, trip helve, crank, steam and air- actuated). The Yoder Company of Cleveland, Ohio, built a double, vertically-configured crank-actuated hammer which will be mentioned briefly later in this volume. The Yoder hammer was, more properly speaking, a lighter-weight, sheet-metal hammer.

[39] Long & Allstatter Company, ca. 1873, p. 22, The original prices were covered with a pasted-in revision (ca. 1880?).

[40] Long & Allstatter Company, *Illustrated Catalogue of Punching and Shearing Machinery and Power Hammers Manufactured by The Long & Allstatter Co.*, Hamilton, Ohio, 1884, pp. 58-59. (Smithsonian Institution). Hammers by both companies were represented by the important New York machinery dealer Manning, Maxwell & Moore. In its catalogue of September, 1884, the 40- and 60-pound hammers of both companies cost the same: $450 and $600 respectively. The 100-pound Long and Allstatter Hammer, at $875, cost $25 more than the Bradley, but the 80-pound Bradley, at $750, cost $75 more than its competitor.

[41] Long and Allstatter Company, *The American Rubber Cushioned Helve Hammer*, Bulletin No. RCH - 1, Hamilton, Ohio, n.d., p. 4. (Smithsonian Institution)

[42] "American Rubber-Cushioned Helve Hammer," *American Machinist*, Vol. 53, No. 3, 1920, p.140.

[43] *The American Rubber Cushioned Helve Hammer*, p. 4.

[44] *Thomas Register of American Manufacturers*, 29th Edition, 1939 (Thomas Publishing Company, New York, Dec., 1938), column 4315.

[45] *American Machinist*, Vol. 7, August 9, 1884, p. 6. Advertisements for the Brown Hammer first appeared in this periodical on Aug. 2 (p. 13).

[46] U. S. Patent no. 327,999 (granted Oct. 13, 1885), p. 1, lines 17-21.

[47] The Knowlton Manufacturing Company was a regular advertiser in *American Machinist* through Sept. of 1887.

[48] *American Machinist*, Vol. 7, Oct. 25, 1884, p. 9.

[49] ibid., Vol. 3, Nov. 6, 1880, p. 5.

[50] U. S Patent no. 206,665 (granted Aug. 6, 1878), p. 2, column 2.

[51] *American Machinist*, Vol. 6, July 7, 1883, p. 9.

[52] ibid, Vol. 7, March 15, p. 12.

[53] Hull & Belden Co., "Shaw & Justice Dead Stroke Power Hammer," brochure, Danbury, early 1875, p.2. (Smithsonian Institution)

[54] "Penney's Spring Cushioned Helve Hammer," *American Machinist*, Vol. 7, March 15, p. 7.

[55] Manning, Maxwell & Moore Co., 1884, p. 296A.

[56] Holt Manufacturing Co., *Bulletin No. MT-22*, Stockton, April, 1905, p. 1. (Carnegie Library, Pittsburgh)

[57] ibid., p. 3.

[58] ibid.

[59] *Thomas' Register of American Manufacturers*, 6th ed., 1914, column 2632.

[60] ibid., column 2638.

[61] "Meet H. W. Hathorn," Mason City newspaper, March 29, 1930.

[62] "New Foundry to be Built," *Mason City Globe-Gazette*, Aug. 27, 1904; and Modern Sales Co. ad., *The American Blacksmith*, Jan., 1913, p. 20.

[63] "Hathorn's Hard Hitting Helve Hammer," *The Blacksmith & Wheelwright*, Vol. 49, Number 2, Feb., 1904, p. 74.

[64] "The Rochester Helve Hammer," *The Blacksmith & Wheelwright*, Nov.,1908, p. 445.

[65] "Meet Henry Hathorn."

[66] West Tire Setter Company, *Rochester Helve Hammer*, brochure, Rochester, N.Y., 1911. (Hagley Library)

[67] West Tire Setter Co. ad., "Blacksmiths! Look at this Rochester Helve Hammer..." *The Blacksmith & Wheelwright*, Nov., 1912, p. 423.

[68] West Tire Setter Co., brochure, p. 3.

[69] "The West Tire Setter Company," *The Blacksmith & Wheelwright*, Jan., 1915, p. 489.

[70] *Thomas' Register of American Manufacturers*, 27th ed., 1937 (Dec., 1936), column 4855.

[71] Bicknell Co. ad., *The Blacksmith & Wheelwright*, Vol. 47, No. 4, April, 1903, pp. 737 and 740.

[72] ibid., Vol. 59, No. 1, Jan., 1909, p. 536.

[73] ibid., Vol. 77. No. 1, Jan., 1918.

[74] ibid.

[75] ibid., Vol. 79 ,No. 3, March, 1919, p. 30.

[76] *Bicknell Manufacturing & Supply Co.*, a brief company profile in the collection of the Rock County Historical Society, Janesville, WI.

[77] "A Home-Made Hammer," *The Blacksmith & Wheelwright*, Vol. 45, No. 1, Jan., 1902, p. 17.

[78] H. Feldhus ad.,*The Blacksmith & Wheelwright*, Vol. 48, No. 2, Aug., 1903, p. 940.

[79] Bloomer Machine Works advertisement, *The Blacksmith & Wheelwright*, Vol. 49, No. 4, April, 1904, p.202.

[80] "Helva-Hard-Hitter Power Hammer," *The Blacksmith & Wheelwright*, Vol. 58, No. 6, June, 1908, p. 252. In assembling these notes, the author discovered a notice in Vol. 51, No. 1, Jan., 1905, p. 609 which states that the company was in the process of "getting out" a small helve hammer "to meet the demand of blacksmiths who do not feel warranted in buying the high grade hammer [the Boss]. It is their opinion [the Novelty Iron Works] that blacksmiths make a mistake in putting in a helve hammer, but if they must have it they are willing to furnish it."

[81] ibid.

[82] *The Blacksmith & Wheelwright*, Vol. 75, No. 4, April, 1917, p. 571.

[83] ibid., Vol. 88, No. 1, p. 16.

[84] ibid. Vol. 69, No. 1, Jan., 1914, p. 29.

[85] ibid. Vol. 101, No. 5, May, 1930, facing page 1.

[86] ibid. Vol. 47, No. 2, Feb., 1903. Ad. states: "50 sold since Sept. 1, 1902."

[87] ibid. Vol. 47, No. 1, Jan., 1903, p. 589.

[88] Richard R. Kern, *The Little Giant Powerhammer* (Xenia, Ohio, H & K Publishing, 1992),p. 174.

[89] Little Giant Co., *Little Giant Power Hammers*, pricesheet, Feb. 1, 1923. (Smithsonian - Heinz Collection)

[90] ibid.

[91] "His Use of Power," *The Blacksmith & Wheelwright*, Vol. 58, No. 11, Nov., 1908, 440.

[92] ibid., Vol. 45, No. 1, Jan., 1902, p. 31.

[93] ibid., Vol. 69, No. 1, Jan., 1914; and Vol. 77, No. 4, April, 1918.

[94] ibid., Vol. 46, No. 2, Aug., 1902, p. 364.

[95] ibid., Vol. 46, No. 6, Dec., 1902, p.540.

[96] ibid., Vol. 47, No. 2, Feb., 1903, p. 634.

[97] ibid., Vol. 54, No. 5, Nov., 1906, p. 496.

[98] ibid., Vol. 54, No. 6, Dec., 1906, p. 546.

[99] ibid., Vol. 61 ,No. 5, June, 1910, p. 240.

[100] *Thomas' Register of American Manufacturers*, 9th ed., Oct., 1917, col. 3613.

[101] *Williams Dayton City Directory*, Williams & Company, Dayton, 1888-1889.

[102] Patterson Tool and Supply Company, trade catalogue, Dayton, Ohio, 1901, p. 654. (Ohio State Historical Society)

[103] "The Dayton Power Hammer," *Modern Machinery*, Jan., 1904, p. 38.

[104] "Dayton Hammer," *Blacksmith and Wheelwright*, Vol. 49, No. 4, April, 1904, pp. 188 and 192.

[105] *Williams Dayton City Directory*, 1903-1904, p. 912.

[106] *Thomas' Register of American Manufacturers*, 4th edition, Aug., 1909, column 955.

[107] ibid., 8th edition, Oct., 1916, column 3313.

[108] "Variable-Blow Power Hammer," *Machinery*, Vol. 17, June, 1911, p. 827.

[109] "Variable Blow Power Hammer," *American Machinist*, May 25, 1911, p. 997.

[110] "Power Hammer," *American Machinist*, May 6, 1909, p. 762.

[111] The story of American foot-powered hammers will be told in another volume. For a study of the origins of the term "Oliver," see Rhys Jenkins, "The Oliver, Iron Making in the Fourteenth Century," *Transactions*, Vol. XII, 1931-1933 (London, The Newcomen Society, 1933), pp. 9-14, plate II.

[112] William Blacker, whose hammer preceeded that of Overcash in England, was also described as a practical blacksmith.

[113] U.S. Patent no. 1,103,918 page 1, line 11.

[114] *The Blacksmith and Wheelwright*, Vol. 71, No. 5, May, 1915, p. 656.

[115] ibid., p. 659.

[116] *The Blacksmith and Wheelwright*, Vol. 73, No. 1, Jan., 1916, p. 41.

[117] The Northwestern Power Hammer Company was never listed in the *Thomas' Register of American Manufacturers*.

[118] *The Blacksmith and Wheelwright*, Vol. 77, No. 4, April, 1918, p. 93.
[119] ibid.
[120] ibid., Vol. 37, No. 3, April, 1898, pp. 60-61.
[121] ibid., Vol. 69, No. 5, May, 1914, pp. 180-181.
[122] ibid., Vol. 77, No. 4, April, 1918, p. 113.
[123] ibid., Vol. 78, No. 3, Sept., 1918, p. 241.
[124] *American Machinist*, Vol. 58, No. 2, Jan. 11, 1923, pp. 91-92.
[125] *The Blacksmith and Wheelwright*, Vol. 87, No. 6, June, 1923, p. 14.
[126] Blacker Engineering Company ,"New 1928 Model" *Blacker Hammers* , trade brochure, 1928. (Chambersburg Engineering Company Archives)
[127] McKiernan-Terry Corporation, "Announcement, Blacker Hammers" (letter), July 5, 1944. (Chambersburg Engineering Company Archives)
[128] *The Anvil's Ring*, Vol. 20, No. 2, Fall, 1992, p. 10 (illustration).
[129] *Machinery*, June, 1926, p. 842.
[130] *Machinery*, Jan., 1928, p.401.
[131] Clarke, Eugene C., *Chambersburg Engineering Company - The First Fifty Years* (Chambersburg, PA., 1948), pp. 45-46.
[132] Howard Terhune, "Blacker Hammers Improved," handwritten internal company memo. (Chambersburg Engineering Company Archives)
[133] Clarke, p. 23.
[134] Apex Forge and Tool Company, 2311 Ross Way, Tacoma, WA, 98421-3402.

* * *

Chapter 4 - *Vertically-configured Guided-ram Hammers*

[1] American Society of Mechanical Engineers, *Mechanical Engineers in America Born Prior to 1861 - A Biographical Dictionary* (New York: A.S.M.E., 1980), pp. 276-277.
[2] "Shaw and Justice's Dead-Stroke Power Hammer," *The Engineer*, Vol. 24, 1867, p. 400.
[3] "Improved Dead-Stroke Hammer," *Scientific American*, Vol. 15, Dec. 8, 1866, p. 383.
[4] ibid.
[5] *Dictionary of American Biography* (New York: Charles Scribner's Sons, 1964), vol. IX, p. 48. The above-cited A.S.M.E. Biographical Dictionary repeats this error.
[6] "Shaw's Gunpowder Piledriver," *Engineering*, Vol. 8, August 6, 1869, pp. 79-80.
[7] "A Gunpowder Hammer," *Engineering* , Vol. 9, Jan. 21, 1870, p. 45. This unusual hammer was granted U.S. patent no. 99,958 on Feb. 15, 1870.
[8] Wilson, Joseph M., *Masterpieces of the Centennial International Exhibition*, Vol. III (Philadelphia, Gebbie & Barrie, 1876), pp. 138-140. See also: "Shaw's

Gunpowder Piledriver," *Engineering*,Vol. 21, May 19, 1876, pp. 408-410.

[9] Philip S. Justice Company, *Testimonials Regarding the Dead-Stroke Power Hammer*, Philadelphia: no date, ca. 1877. (Hagley Library)

[10] ibid.

[11] "Dead Stroke Power Hammer," *The Technologist, or Industrial Monthly* (January, 1876), p.12.

[12] "Power-Hammer," United States Patent Office, Patent no. 403,453 (granted May 14, 1889, to Russell A. Belden of New Haven, Connecticut), lines 17-19.

[13] Wilson, p. 116.

[14] *American Machinist*, Vol. 11, Sept. 1, 1888, p. 9.

[15] *The Iron Age* , Vol. 42, Oct. 4, 1888, p. 507.

[16] "Upright Power Hammer," *The Iron Age*, Vol. 43, 1889, p. 394.

[17] Williams, White and Company, *The First Hundred Years* (Moline,1954), p. 36.

[18] Williams, White and Company, *Machinery Catalogue No. 8*, Moline, no date, p. 62. (Carnegie Library, Pittsburg)

[19] ibid.

[20] Williams, White and Company, *Machinery Catalogue No. 10*, Moline, no date, pp. 75-76. (Chambersburg Engineering Company archives)

[21] "Power-Hammer," United States Patent Office, Patent no. 244,927 (granted July 26, 1881, John Patterson, of London), lines 28-61.

[22] *The Industries of Philadelphia* (Philadelphia: Richard Edwards, 1881), p. 134.

[23] Manning, Maxwell and Moore Co., *Illustrated Catalogue of Railway and Machinists' Tools and Supplies* (New York: Jan., 1894), p. 287. (Stanford University Library)

[24] Barton, Timothy, "Panel Beaters Round-up," *Street Rodder* (Anaheim, California), Vol. 21, No. 8, Aug., 1992, pp. 136-143.

[25] W. B. Gay, *Gazetteer - Hampshire County, Mass., 1654-1887, Part First* (Syracuse, N. Y., W. B. Gay & Co., 1887), p. 365.

[26] The Howe Scale Company, *Howe Power Hammer*, pamphlet, Rutland, Vermont, no date. (Bailey-Howe Library, University of Vermont, Burlington.

[27] Fairbanks, Edward T., *The Town of St. Johnsbury VT* (St. Johnsbury: the Cowles Press, 1914), p. 476.

[28] *Caledonian* (St. Johnsbury, Vt.), Feb. 12, 1891.

[29] ibid.

[30] *Caledonian*, March 31, 1892.

[31] "Fairbanks Power Hammers," *The Blacksmith and Wheelwright*, Vol. 75, No. 1, Jan., 1917, p. 479.

[32] *St. Johnsbury City Directory*, 1909.

[33] "Power Hammer," *American Machinist* , Vol. 39, No. 14, 1913, p. 586.

[34] " 'Fairbanks' Power Hammers With New Gib and Faceplate," *American Machinist*, Vol. 48, No. 14, p. 598.

[35] "Guards, Safety, for Power Hammers," Condensed-Clipping Index of Equipment, *American Machinist*, vol. 52, 1920, p. 110c.

[36] Norris, Kimberly (purchasing agent for Barbour-Stockwell Company), letter of June 17, 1994.

[37] Beaudry and Company, *The Beaudry Power Hammers* (Boston: no date, but ca. 1912), p. 4. (Carnegie Library, Pittsburg)

[38] ibid., p. 7.

[39] ibid., p. 3.

[40] Beaudry and Company, *Beaudry Hammers* (Boston: no date), pp. 8-10. (Smithsonian Institution) See also: "Beaudry Utility Hammer," *American Machinist*, Vol. 61, 1924, p. 827.

[41] "Beaudry Air-Cushioned Helve Hammer," *American Machinist*, Vol. 61, 1924, p. 1016.

[42] "Beaudry Upright Air Hammer," *American Machinist*, Vol. 62, No. 4, Jan. 22, 1925, p. 180.

[43] *American Machinist* , Vol. 70, No. 12, 1929, p. 496d.

[44] *American Machinist* , Vol. 71, Oct. 3, 1929, p. 610g.

[45] "The Bradley Compact Hammer," *The Iron Age*, Oct. 5, 1899, pp. 1-2.

[46] Little Giant, Inc., *Price and Parts List*, Jan. 1, 1942. (Smithsonian Institution)

[47] Kane & Roach, *Catalogue No. 6 - Carriage & Wagon Machinery*, Syracuse, 1919, p. 46. (Smithsonian Institution)

[48] Kane & Roach, *Fifty Years of Service*, Syracuse, 1937, Foreword.

[49] *Thomas' Register of American Manufacturers*, 35th ed., Dec., 1944, column 6294.

[50] The Hotchkiss hammer is described in greater detail in the section of this book devoted to atmospheric hammers.

[51] The application for Melvin Yeakley's first patent, no. 568, 395 was filed Oct. 11, 1895. It was granted Sept. 29, 1896. Production of the hammer by Williams, White & Company, of Moline, Illinois, began in 1898 [*The First Hundred Years* (Williams, White & Company, Moline, 1954) p. 47]. Earlier introduction of the hammer by Yeakley himself is quite likely.

[52] U.S. Patent no. 485,498, p. 2, lines 112-120.

[53] *American Machinist*, Vol. 11, No. 36, Sept. 8, 1888, p.1.

[54] *The Engineer*, Vol. 66, 1888, pp. 79-80.

[55] *Annewalt's Cleveland Directory*, Vols. 18 &19, 1889 & 1890.

[56] *American Machinist*, Vol. 14, April 30, 1891.
[57] ibid., May 14, 1891, p. 1.
[58] *The Iron Age*, Vol. 48, July 2, 1891, p. 22.
[59] ibid., Vol. 49, May 12, 1892, p. 922.
[60] E.A. Kinsey & Co. catalogue, Cincinatti, Ohio, no date, p. 527-529. (Ohio Historical Society)
[61] Manning, Maxwell & Moore Co., Jan.,1894, pp. 285-288. (Stanford University Library)
[62] *Annewalt's Cleveland Directory*, Vol. 22, 1892.
[63] ibid., Vol.25, 1895.
[64] Kern, p. 164.
[65] ibid., p. 137.
[66] *The Blacksmith and Wheelwright*, Vol. 93, No. 1, Jan., 1926, p. 10.
[67] Kern, p. 164.
[68] *The Blacksmith and Wheelwright*, Vol. 39, No. 4, April, 1899, p. 314.
[69] Little Giant Company, *Little Giant Power Hammers* (price list, 1924) includes the neckyoke singletree hammer, a list from 1926 does not. (Smithsonian Institution)
[70] Little Giant Company, *Little Giant Power Hammers* (price list, 1923). The 1,000 pound hammer is designated model "V." (Smithsonian)
[71] Kern, p. 153.
[72] "Mayer Power Hammers," *American Machinist*, Vol. 52, Jan. 15, 1920, p. 159.
[73] "Joy's Power Hammer," *Engineering*, Vol. 7, Oct. 29, 1869, p. 294.
[74] Little Giant Company, price list, 1923.
[75] *American Machinist* , Vol. 70, 1929, p. 74f. The author has a haunting recollection that he has seen a hammer with "Brazil, Indiana" on the crankplate, but cannot verify it...
[76] D.J. Murray Manufacturing Company, letter of April 15, 1939, from C.E. Staky to Herberts-Moore Machinery Company, San Francisco. (Smithsonian)
[77] Wodalski, Ed, "Murray's roots run to lumber era," *Wausau Sunday Herald*, Nov. 16, 1986, p.8c. Also: Pirkl, Tim, "Murray demolishes part of landmark building," *Wausau Daily Herald*, May 10, 1989.
[78] Harlan "Sid" Suedmeier can be contacted through Little Giant Automotive, Inc., 420 4th Corso, Nebraska City, Nebraska, 68410.
[79] *The Blacksmith & Wheelwright*, Vol. 55, No. 6, June, 1907, p. 811
[80] "A Good Machine and the Best Advice," *The Blacksmith & Wheelwright*, Vol. 61, No. 5, June, 1910, p. 226.
[81] "A Good Power Hammer," *The Blacksmith & Wheelwright*, Vol. 59, No. 3, Mar., 1909, p. 607.

82 "A Home-Made Hammer," *The Blacksmith & Wheelwright*, Vol. 45, No. 1, Jan., 1902, p. 17.

83 *The Blacksmith & Wheelwright*, Vol. 49, No. 1, Jan., 1904, p. 40. The crank plate named Feldhus as the manufacturer, The W. J. Broatch Iron Co. of Omaha, Nebraska was listed as sole agent. By Jan. of 1906, the Little John Power Hammer was being offered by the Lincoln Mfg. and Dev. Co. of Lincoln: Vol. 53, No. 1., p. 34.

84 ibid., Vol. 50, No. 6, Dec., 1904, p. 531. (see Vol. 49, No. 5, May, 1904, p. 240 for possibly an earlier listing)

85 "The Hilbert-Freiberg Power Post Hammer," *The Iron Age*, May 31, 1900, p. 13.

86 *The Blacksmith & Wheelwright*, Vol 45, No. 3, March, 1902, p. 122.

87 "North Reading's Industrial Past - The McLane Wagon Works," *North Reading Review*, No. 5, Dec., 1963. p. 17-19.

88 *The Blacksmith & Wheelwright*, Vol. 58, No. 8, Aug., 1908, p. 309.

89 Champion Blower & Forge Co., brochure, Jan. 1, 1912, p. 34. (author's library)

90 Champion Blower & Forge Co., *Illustrated Catalogue, January, 1913*, p. 201. (Carnegie Library, Pittsburg)

91 Champion Blower & Forge Co., *Illustrated Catalogue, January, 1920*, p. 208-211 (author's library)

92 Champion Blower & Forge Co., *Catalogue No. 60*, n.d., ca. 1936, p. 64. (Smithsonian)

93 *The Blacksmith & Wheelwright*, Vol. 51, No. 3, March, 1905, pp. 688 & 691.

94 ibid., Vol. 63, No. 3, March, 1911.

95 ibid., Vol. 67, No. 4, April, 1913, p. 623.

96 State of Texas, *Charter of Incorporation* (No. 46737), Sept. 15, 1926.

97 ibid., Vol. 95, No. 1, Jan., 1927, p. ?; also, May, p. 13.

98 Berger & Carter Co. catalogue, San Francisco, 1918, pp. 344-346. (California Historical Society, San Francisco)

99 *The Blacksmith & Wheelwright*, Vol. 50, No. 1, July, 1904, p. 303. Interestingly, George E. Davis, of the Novelty Iron Works, announced that there would be a harrow disc sharpening attachment for the Boss hammer, but it is not known if this device was ever placed on the market (See Vol. 51, No. 21, Jan., 1905, p. 609).

100 "Dies for Plow Work," *The Blacksmith & Wheelwright*, Vol. 61, No. 1, Jan., 1910, p. 32.

101 *The Blacksmith & Wheelwright*, Vol. 49, No. 1, Jan., 1904, p. 42.

102 ibid., Vol. 51, No. 1, Jan., 1905, p. 614.

103 U.S. patent 726,879 (granted May 5, 1903), p. 1, lines 12-14.

[104] *The Blacksmith & Wheelwright*, Vol. 55, No. 1, Jan., 1907, p. 586.

[105] ibid., Vol. 45, No. 2, Feb., 1902, p. 68.

[106] ibid. Vol. 48, No. 6, Dec., 1903, p. 1143.

[107] ibid., Vol. 57, No. 1, Jan., 1908, p. 30.

[108] ibid., Vol. 55, No. 1, Jan., 1907, p. 586.

[109] ibid., Vol. 59, No. 3, March, 1909, p. 619.

[110] ibid., Vol. 67, No. 4, April, 1913, p. 598.

[111] ibid., Vol. 75, No. 4, April, 1917, p. 571. This version was known as the New Idea Power Hammer. See also, Vol. 65, No. 1, Jan., 1912.

[112] *The Blacksmith & Wheelwright*, Vol. 88, No. 1, June, 1923, p. 16.

[113] ibid., Vol. 49, No. 2, Feb., 1904, p. 87.

[114] ibid., Vol. 52, No.1, July, 1905, p. 846

[115] ibid., p. 855

[116] ibid., Vol. 55, No. 2, Feb., 1907, p. 633

[117] Minneapolis Iron Store Co., *Illustrated Catalogue of Standard Heavy Hardware*, 1906, pp. 192-193. (Minneapolis Public Library)

[118] *Thomas' Register of American Manufacturers*, 8th ed., Oct., 1916, col. 3318.

[119] *The Towne Gossipe*, November, 1923, p. 17.

[120] *Massillon City Directory*, 1920, p. 218.

[121] *Thomas' Register of American Manufacturers*, 19th ed., Aug., 1928, col. 4257.

[122] Waterhouse & Lester Co., *Catalogue No. 20*, San Francisco, 1924, p. 336.

[123] *Bolts & Nuts*, newsletter of the manufacturers' committee, Oakland Chamber of Commerce, June 12, 1922.

[124] Hendrie & Boltoff Manufacturing & Supply Co., *Catalogue No. 48*, Denver, 1917, p. 328. (author's library)

[125] Pat Narver (granddaughter of Henry E. Brett), letter of March 19, 1994.

[126] *Machinery*, Vol. 16, March, 1910, p. 595.

[127] Joseph T. Ryerson & Son, *Bulletin* 8,905, Chicago, n.d.., pp. 8,905-8,912. (Carnegie Library, Pittsburgh)

[128] ibid, p. 8,026 (for Little Giant). See also: Joseph T. Ryerson, *Ryerson Reference Book*, Chicago, 1910, pp. 266-269 (for Modern and Hathorn helve).

[129] *The Blacksmith & Wheelwright*, Vol. 55, No. 2, Feb., 1907, p. 622.

[130] ibid., Vol. 59, No. 7, July, 1909, p. 761.

[131] *The Blacksmith & Wheelwright*, Vol. 53, No. 6, June, 1906. p. 266.

* * *

Chapter 5 - *Non-vertically-configured Guided-ram Hammers*

[1] S.C. Forsaith & Co., *The Palmer Power Spring Hammer*, brochure/price sheet, ca. 1874, p. 2. (Smithsonian)

[2] ibid.

[3] "Palmer Power Spring Hammer," *Scientific American*, Oct. 25, 1873, p. 262.

[4] S.C. Forsaith & Co., *The Palmer Power Spring Hammer.*

[5] ibid., p. 3.

[6] ibid.

[7] *The Concord Directory for 1880-81* (Boston, Greenough & Co., 1880), p. 8.

[8] "Dunham's Spring Beam Hammer," *Engineering*, Vol. 32, Nov. 11, 1881, p. 485.

[9] *Leading Manufacturers and Merchants of New Hampshire* (New York, International Publishing Co., 1887), p. 160.

[10] "Samuel Caldwell Forsaith, who invented a machine for folding newspapers," newspaper profile from the files of the Manchester Historic Association, (source unknown).

[11] *The Blacksmith & Wheelwright*, Vol. 7, No. 6, June, 1883, p. 214.

[12] *Manchester City Directory, 1897*, p. 643 (advertisement).

[13] State of New Hampshire, *Chapter 333 - An Act to Repeal Charters of Certain Inactive and Delinquent Corporations* , 1925, pp. 328-329.

[14] *Machinery*, Vol. 15, March, 1909, p. 570 (an obituary notice for Alexander Beaudry).

[15] *American Machinist*, Vol. 8, June 6, 1885, p. 10.

[16] ibid., Vol. 9, Oct. 23, 1886, p. 15.

[17] ibid., Vol. 9, Jan. 16, 1886, p. 2.

[18] Advertisements from 1885 already note that "they are being used extensively by cutlery manufacturers and wagon makers." The patent reference is to figure 1. Figure 3, a cross- section is illustrated here, for its clarity and detail. For a profile of Beaudry's Upright Cushioned Hammer, see *American Machinist*, Vol. 9, Feb. 6, 1886, p. 7.

[19] Advertisements listing the new name first appear in *American Machinist*, Vol. 10, May 14, 1887, p. 11.

[20] *American Machinist*, Vol. 11, No. 2, Jan. 14, 1888, p. 9.

[21] ibid., Vol. 12, No. 1, Jan. 3, 1889, p. 7.

[22] ibid., Vol. 12, No. 17, April 25, 1889, p. 1.

[23] U.S. Patent no. 461,917, p. 1, column 1, lines 9-11.

[24] *American Machinist*, Vol. 13, No. 22, May 29, 1890, p. 11.

[25] *Iron Age*, Vol. 49, Jan. 21, 1892, p. 114.

[26] *Machinery*, Nov., 1924, p. 254.

[27] *American Machinist*, Vol. 61, No. 26, 1924, p. 1016.

[28] *Machinery*, Feb., 1925, p. 493.

[29] *American Machinist*, Vol. 12, No. 19, May 9, 1889, p. 10.

[30] ibid., Vol. 8, Sept. 26, 1885, p. 7.

[31] C.C. Bradley & Son, Inc., *The Bradley Upright Helve Hammer*, brochure, 1920, p. 1. (Chambersburg Engineering Company archives)

[32] C.C. Bradley & Son, Inc., *The Bradley Upright Strap Hammer*, brochure,1920, p.2 (Chambersburg Engineering Company archives)

[33] "The Bradley Hammer," *The Iron Age*, Vol. 45, Feb., 1890, p. 291.

[34] *The Bradley Upright Strap Hammer.*

[35] Robert L. Merriam, et al., *The History of the John Russell Cutlery Company* Greenfield, Mass., Bete Press, 1976), p. 30.

[36] C.C. Bradley & Son, Inc., *The New Bradley Rubber Cushioned Helve Hammer* , *The Bradley Upright Strap Hammer*, and *The New Bradley Compact Hammer*, brochures/price sheets, Nov. 1, 1939. (Carnegie Library, Pittsburgh)

[37] U. S. patent no. 156,276 (granted Oct. 27, 1874), p. 1.

[38] *Democratic Watchman* (Bellefonte, PA), April 17, 1908.

[39] John Blair Linn, *History of Centre and Clinton Counties, Pennsylvania* (Philadelphia, Louis H. Everts, 1883), p. 244.

[40] W. P. Duncan & Company, *Vulcan Power Hammer*, Bellefonte, 1885, p. 3. (Hagley Library)

[41] Manning, Maxwell & Moore Co., January, 1894, p. 286.

[42] Jenkins & Lingle Co., *Price List of Jenkins' Cushioned Power Hammer* Bellefonte, n.d.. (Hagley Library)

[43] *Vulcan Power Hammer*, p. 2-3.

[44] C. Chomienne, " The Duncan Helve-Hammer" (Chapter XXXVIII) from "Notes On Steam Hammers" *The Railroad and Engineering Journal*, Vol. 58, No. 3, March, 1889, p. 126.

[45] *Iron Age*, Vol. 47, March 26, 1891, p. 585.

[46] Jenkins & Lingle Co., *Jenkins' Upright Cushioned Helve Hammer*, Bellefonte, n.d.), p. 3. (Hagley Library)

[47] *Democratic Watchman.*

[48] *The Blacksmith & Wheelwright*, Vol. 47, No. 3, March, 1903, p.682.

[49] Longworth, Daniel, "On Power Hammers with a Movable Fulcrum," *Proceedings*, Institution of Mechanical Engineers (London), 1882, pp. 204-213. For an early profile of Longworth's design, see "Longworth's Patent Forging Hammers and Stamps," *The Engineer*, Vol. 44, 1877, p.4.

[50] J. Scott Rawlings was soon succeeded by W. & J. Player.

[51] *Proceedings*, plate 29.

[52] "Longworth's Patent Power Hammer," *The Engineer*, Vol. 73, Feb. 26, 1892, p. 177.

[53] ibid.

[54] Samuelson, Ernest, "Improvements in the Longworth Power-Hammer," *Proceedings*, Institution of Mechanical Engineers (London), 1900, pp. 99-118.

[55] *Montezuma Weekly Republican*, March 11, 1903, p.1.

[56] *Grinnell-Herald*, Aug. 28 and Sept. 8 (supplement), 1903.

[57] Curt McConnell, *Great Cars of the Great Plains* (Lincoln, U. of Nebraska Press, 1995), pp. 135-136.

[58] *Grinnell-Herald*, Nov. 27, 1903, p. 3.

[59] ibid., Dec. 11, 1904, p. 9.

[60] ibid., Feb. 23, 1904, p. 5.

[61] ibid., July 5, 1904, p. 3.

[62] *Mason City Globe Gazette*, August 27, 1904.

[63] *History of Poweshiek County* (1911), Vol. 2, p. 290.

[64] *Grinnell-Herald*, Aug. 11, 1908, p. 1.

[65] ibid., Sept 1, 1908, p. 2.

[66] ibid., July 23, 1909, p. 1.

[67] ibid., Oct. 8, 1909, p. 1.

[68] ibid., Aug. 19, 1910, p. 3.

[69] *The American Blacksmith*, Jan., 1913, p. 20.

[70] *The Blacksmith & Wheelwright*, Vol. 68, No. 6, Dec., 1913, p. 903.

[71] *The Powerhammer*, Vol. 1, No. 3, May/June, 1992, pp. 9-10.

[72] Alinder Hammer Co., *The Alinder Power Hammer* Milwaukee, November, 1917), p.4. (Chambersburg Engineering Company archives)

[73] ibid., p. 3.

[74] ibid.

[75] *Machinery*, Vol. 17, Jan., 1911, p. 422.

[76] *Wisconsin State Gazetteer and Business Directory for 1911-12* (R. L. Polk & Co.), p. 888.

[77] An entry was found for "Alinder, Hen." in an employee ledger from Allis-Chalmers dated "9/06." This ledger is part of the Allis-Chalmers manuscript collection held by the Milwaukee County Historical Society. Subsequent intermittent listings for Henry Alinder as a draftsman at Allis-Chalmers are to be found in Milwaukee City Directories.

[78] *History of Chippewa County* (1913), p. 209.

[79] Thomas W. Enderson, *The Keller Engine* (1994, privately-printed), p.2. Mr. Enderson may be contacted at RR1 Box 227, Jim Falls, WI, 54748.

[80] J. E. Keller moved to Menomenie, Wisconsin, in 1905, to work for the Oscillating Sleigh Company as a blacksmith. He remained there until 1916, at which time he established his own blacksmithing and repair business (Enderson, p. 6).

[81] Enderson, p. 12.

[82] *Thomas' Register of American Manufacturers*, 9th ed., Oct., 1917, column. 3620.

[83] *Post* (New Ulm), April 1, 1870.

[84] ibid., May 23, 1879.

[85] *Review* (New Ulm), March 11, 1896.

[86] "An Upright Power Hammer," *American Machinist*, Vol. 30, February 7, 1907, p. 200.

[87] *Review*, March 30, 1910.

[88] *Thomas' Register Of American Manufacturers*, 14th ed., 1923-1924, col. 4074.

[89] U.S. patent no. 412,809 (granted Oct. 15, 1889), p. 1, lines 15-16

[90] "Vaughn's Patent Power Hammer," *American Machinist*, Oct. 8, 1891.

[91] U.S. patent no. 606,370 (granted June 28, 1898), p. 1, lines 30-34.

[92] "New Power Hammer," *The Journal of Railway Appliances*," June, 1897, p. 187.

[93] *Worchester Directory*, 1915.

[94] *Thomas' Register of American Manufacturers*, 7th ed., Oct. 1915, col. 2974.

[95] ibid., 5th ed., 1912, col. 2378.

[96] *The Blacksmith & Wheelwright*, Vol. 88, No. 6, Dec., 1923, p. 2.

[97] Articles of Incorporation were signed Jan. 20, 1900; capital stock: $10,000.

[98] *Minneapolis City Directory*, 1923, p. 456 (advertisement).

[99] John "Ken" Monson, letter of Sept. 7, 1996.

[100] *Certificate of Amendment of Article of Incorporation*, filed Dec. 11, 1968.

[101] *Consent to the Use of an Assumed Name or Corporate Name*, filed Jan. 25, 1984.

[102] *Certificate of Involuntary Dissolution*, dated Sept. 27, 1991.

[103] Paul Lague, president, TC/American Monorail, letter of Jan. 15, 1997. This company has no involvement with power hammers, and has no specific information about the Central Power Hammer.

[104] *Machinery*, Vol. 4, No. 6, Feb., 1898, p.38.

[105] *Blacksmith & Wheelwright*, Vol. 59, No. 2, Feb., 1909, p.575.

[106] Cray Brothers, *Net Price Catalogue* (Cray Brothers, Cleveland, Ohio, 1909), p. 10.

[107] Red Oak Commercial Club, *Red Oak at the Dawn of the Twentieth Century* (Red Oak Express,Red Oak, 1900).

[108] *Red Oak Independent*, Christmas Issue, 1894, p. 14.

[109] Red Oak Commercial Club.

[110] *Red Oak Sun*, November 3, 1901.

[111] ibid., April 12, 1907.

[112] Kerrihard Co. brochure (Red Oak, n.d., ca. 1918), pp. 10-20. (author's library)

[113] *Blacksmith & Wheelwright*, Vol. 101, No. 5, May, 1930.

[114] ibid.

[115] Montgomery County Centennial Corp., *Cavalcade of a Century - Official Program* (Business and Professional Women's Club, Red Oak, Aug., 1953), p. 48.

[116] "A Power Hammer," *American Machinist*, July 20, 1905, p.98.

[117] *Twentieth Century Directory of the Cities of Sterling & Rock Falls, Illinois* (Sterling, Werntz & Hubbard, 1901), p. 68.

[118] *Directory of the Cities of Sterling & Rock Falls, 1904* (Sterling, Charles W. Brion, 1904), p. 130.

[119] *Directory of the Cities of Sterling & Rock Falls, 1906* (Sterling, W. Millard Haskell, 1906), p. 159.

[120] "The First Automobiles," *Sterling Gazette*, June 6, 1909.

[121] *The Blacksmith & Wheelwright*, Vol. 61, No. 5, June, 1910, p. 240.

[122] *Thomas' Register of American Manufacturers*, 5th ed., 1912, col. 2378.

[123] U.S. patent no. 810,905 (granted Jan. 30, 1906), p. 2, lines 8-9.

[124] ibid., p. 1, lines 112-113.

[125] "A New Power Hammer," *American Machinist*, Sept. 10, 1903, p. 1307.

[126] "New Power Hammer," *American Machinist*, April 14, 1910, p. 711.

[127] U.S. patent 477,159 (granted June 14, 1892, p. 2, line 71.

[128] "The Laird & Sweeney Power Hammer, " *The Iron Age*, Vol. 51, Feb. 9, 1893, pp. 307-308.

[129] Edward T. Fairbanks, pp. 477-478.

* * *

Chapter 6 - *Concluding Remarks*

[1] Douglas T. Hamilton & Erik Oberg, *Electric Welding* (New York, The Industrial Press, 1918), pp. 9-16.

[2] "A Review of Progress in Arc Welding," *The Iron Age*, Vol. 126, No. 21, Nov. 20, 1930, p. 512.

[3] Robert E. Smith, *Forging & Welding* (Bloomington, Illinois, McKnight & McKnight Publishing, 1956, pp. 39-40.

[4] The American Society of Mechanical Engineers, *Condensed Catalogues of Mechanical Equipment*, 1927-28, 17th annual ed. (New York, Oct. 1927), p. 584.

* * *

Bibliography

Periodicals

Anvil's Ring (Artist-Blacksmiths' Association of North America)
American Blacksmith
American Machinist
The Blacksmith & Wheelwright
The Chronicle (Early American Industries Association)
Condensed Catalogues of Mechanical Equipment (A.S.M.E.)
Dr. Dingler's Polytechnischen Journal
The Engineer
Engineering
The Iron Age
Journal of Railway Appliances and Railway Price Current
Machinery
The Powerhammer
The Railroad and Engineering Journal
The Scientific American
The Technologist, or Industrial Monthly
Thomas' Register of American Manufacturers

Trade Literature & Local Sources

Company catalogues, brochures, and price sheets are described in the appropriate notes for each chapter (in many cases the location of these rare materials is provided to assist the reader in further research. Books are generally easy to locate, but trade literature is another matter.) City directories and local newspapers for the places in which hammer-building firms were situated are also mentioned in the notes.

Books

American Society of Mechanical Engineers: the History & Heritage Committee, *Mechanical Engineers in America Born Prior to 1861 - A Biographical Dictionary* (New York, American Soc. of Mechanical Engineers, 1980).

Bealer, Alex W., *The Art of the Blacksmith* (New York, Funk and Wagnall's, 1969)

Cantrell, J.A., *James Nasmyth and the Bridgewater Foundry - A study of entrepreneur ship in the early engineering industry* (Manchester, Chetham Soc., 1984).

Caterpillar Tractor Company, *Fifty Years on Tracks* (Peoria, IL., Caterpillar Tractor Company, 1954).

Chambersburg Engineering Company, *Chambersburg Engineering Company - The First Fifty Years, 1897-1947* (Chambersburg, Chambersburg Engineering Company, 1947).

Clemens, A.B., *Forging and Tool Dressing* (Scranton, International Textbook Company, 1924).

Cox, Tench, *A Statement of the Arts and Manufacture of the United States for the year 1810* (Philadelphia, U. S. Treasury Department, 1814).

Enderson, Thomas, *The Keller Engine* (Jim Falls, privately printed, 1994).

Fairbanks, Edward T., *The Town of St. Johnsbury VT* (St. Johnsbury, Cowles Press, 1914).

Flink, James J., *America Adopts the Automobile* (Cambridge, the MIT Press, 1970).

Flohr, Carl (Company), *Carl-Flohr - Fest-Schrift zur Feier des 25 Jahrigen Geschaftsjubilaums* (Berlin, Carl Flohr, 1904).

Hamilton, Douglas T., and Oberg, Eric, *Electric Welding* (New York, The Industrial Press, 1918).

Holstrom, J.G., *Drake's Modern Blacksmithing and Horseshoeing* (New York, Drake Publishers, 1971).

Hurt, R. Douglas, *American Farm Tools - from Hand-power to Steam-power* (Manhattan, Kansas, Sunflower University Press, 1982).

Hutton, F. R., *Report on Machine-tools and Wood-working Machinery* (Washington, Department of the Interior - Census Office, 1885).

International Library of Technology, *Machine Moulding...Special Forging Operations* (Scranton, International Textbook Company, 1906).

International Publishing Company, *Leading Manufacturers and merchants of New Hampshire* (New York, International Publishing Company, 1887).

Jenkins, Rhys, "The Oliver. - Iron Making in the Fourteenth Century," *Transactions Vol. XII, 1931-1932* (London, The Newcomen Soc., 1933).

Kane & Roach (Company), *Fifty Years of Service* (Syracuse, New York, Kane & Roach, 1937).

Kern, Richard R., *The Little Giant Powerhammer* (Xenia, OH, H & K Publishing, 1992).

McConnell, Curt, *Great Cars of the Great Plains* (Lincoln, University of Nebraska Press, 1995).

Merriam, Robert L., et al., *The History of the John Russell Cutlery Company* (Greenfield, Bete Press, 1976).

National Publishing Company, *Commerce, Manufacturers & Resources of New Haven, Conn. - A Historical, Statistical and Descriptive Review* (New Haven, National Publishing Company, 1882).

Naujoks, Waldemar and Fabel, Donald C., *Forging Handbook* (Cleveland, American Society for Metals, 1939).

Nicholson File Company, *File Filosophy* (Providence, R.I., Nicholson File Company, 1943).

Office of Price Administration, *Prices of New Machine Tools as of March 1, 1941* (Washington D.C., Office of Price Administration, 1941).

Open Die Forging Institute, *Open Die Forging Manual* (New York, Open Die Forging Institute, 1949).

Overman, Frederick, *The Manufacture of Iron in All its Various Branches* (Philadelphia, Henry C. Baird,1851).

Reichelt, Richard, *Heartland Blacksmiths - Conversations at the Forge* (Carbondale, Southern Illinois University Press, 1988).

Richardson, M.T., *Practical Blacksmithing* (New York, M.T. Richardson, 1889-1891).

Rose, Joshua, *Modern Machine Shop Practice* (New York, Charles Scribner's Sons, 1887).

Rosenberg, Nathan, ed., *The American System of Manufactures - Report of the Committee on the Machinery of the United States, 1855* (Edinburgh, University of Edinburgh Press, 1969).

Roe, Joseph Wickam, *English and American Tool Builders* (New Haven, Yale University Press, 1916).

Scott, John, *Genius Rewarded; or the Story of the Sewing Machine* (New York, John J. Caulon, 1880).

Smiles, Samuel, ed. *James Nasmyth - an Autobiography* (New York, Harper and Brothers, 1883).

Smith, H.R. Bradley, *Blacksmiths' and Farriers' Tools at the Shelbourne Museum - A History of their Development from Forge to Factory* (Shelbourne, The Shelbourne Museum, 1966).

Smith, Robert E., *Forging and Welding* (Bloomington, Illinois, McKnight & McKnight Publishing, 1956).

Thomas, John J., *Farm Implements and Farm Machinery* (New York, Orange Judd & Company, 1869).

Morrell, Daniel J., et al, *Reports of the United States Commissioners to the Paris Universal Exposition - 1878 - Volume III* (Washington D.C., Government Printing Office, 1880).

Wendel, C. H., *American Gasoline Engines Since 1872* (Sarasota, Crestline Publishing, 1983).

Williams-White & Company, *The First Hundred Years 1854-1954* (Moline, Williams-White & Company, 1954).

Wilson, Joseph M., *The Masterpieces of the Centennial International Exhibition, Illustrated*; Volume III (Philadelphia, Gebbie & Barrie, 187?).

Index

1909 Improved Power Hammer, 267-269
Abrahamsen, Otto, 157
actuation, (term defined) 18-19
Aktiebolaget Klafrestroms Druk, 254
Alinder Hammer, 251-254
Alinder Hammer Co., 253
Alinder, Henry, 212,251-254,**252**,273
Allen, G.W., 179
Allen, J.R., 111
American Fork & Hoe Co., 73
American Rubber Cushioned Helve Hammer, 78-81,**79-80**
Amoskeag Mill machine shop, 216
Apex Forge & Tool Co., 132
Armstrong Manufacturing Co., 102, 195
axles, welding, 22, 28-29,283
Barbour-Stockwell Co., 152
Beaudry, Alexander, 112,113,216, 217,**217**,**219**,**221-222**,225, 241
Beaudry, Augustin, 153-**154**,223
Beaudry & Co., 5,70,73,112,153-157, 169,170,210,216-225,236, 280
Beaudry & Cunningham, 216-219
Beaudry Champion Power Hammer, 153-157,**153**
Beaudry Cylinder Power Hammer, 221-223,**222**
Beaudry Peerless Power Hammer, 155-156
Beaudry Pillar Hammer, **220**-221
Beaudry Upright Cushioned Hammer, 217,221,**218**,**230**
Beaudry Utility Power Hammer, 156-157
Belden Improved Power Hammer, 142
Belden, Russell, 141,143,**142**,149, 168,191,211,225
Belden Machine Co., 142-143
Bement-Miles Co., 130
Bethlehem Steel, 17
Bicknell Manufacturing & Supply Co., 98-99
Bicknell's Power Hammer, **98**-99
Bicknell's Triple Helve Hammer, **99**
Bierhart, Henry, 29,70
Billings & Spenser Co., 35
Blacker Engineering Co., 128-130
Blacker Hammer, **21**,116,121-130, **122**,131-132,161
Blacker Model "B," 128-130
Blacker Hammer Co., 128
Blacker, William, 21,115,121-130, **123**,**129**,131
Blacksmith's Friend Power Hammer, **100**,106
Bloomer Machine Works, 101,254-257,**255**
Blossom, Edward H., 151-**152**
Boss Power Hammer, 29,101-102, 120,192-196,**193-196**,199
bracket (post) hammers, 147-**148**, 181-**182**
Bradley, Charles Goodwin, 74

Bradley, Christopher C. (several), 61-76,**70**,150,229
Bradley & Co., 7,29,53,60-76,**62-74**,90,95,142,144,150,157,169,219,223-233
Bradley Ball Bearing Shaft Coupler, 70-**71**
Bradley Compact Hammer, 70,157-159,**158**,231,233,270
Bradley-Edlund Corp., 74-75
Bradley Hard Coal Heating Forge, **70**
Bradley Rubber Cushioned Helve Hammer, 18, 60-76,**63**,**66**,**68**,79,81,92,95,136,223,225,231-232,238
Bradley Upright Helve Hammer, 18, 223-227,**226**,257
Bradley Upright (Cushioned) Strap Hammer, **20**,72,83,157,168,223-231,**228**,238,259-260
Bragg, Austin, 273-275,**274**
Bragg's Hammer, **273**-275
Brett, Henry E., 204-205
Brooksbank, Fredrick C., 161-168
Brown, Charles M., 81-86,**82**,**84-85**,230
Brown, James A., 204-**205**
Brown Slow Raising Quick Striking Hammer, 81-86, **82**,**85**,231
Butterfield, John C., **60**,62-65,90,210,231-236,**232-235**
C.C. Bradley & Sons, SEE Bradley & Co.
C-30 Power Hammer, 269-**270**
Carr, Charles J., 39
Central Machine Works Co., 263-265
Central Power Hammer, 263-265,**264**
Chambersburg Engineering Co., 121,130-132
Chambersburg Mechanical Sledge Hammer, 121,128,130-132,**131**,132
Champion Blower & Forge Co., 182-187,**185-186**
Champion "Patented" Power Hammer, 182-187,**186**
Charles Merrill & Sons, 47,50-51
Chenot's Atmospheric Hammer, 52
Christian, Elroy, 16
Clark Olds & Co., 207
Clock, E.E., 249-250
closed-die, (term defined) 8-11,34
coil-spring linkage, 45
Coleman Iron Works, 197
Cortland Little Giant Power Hammer, **196**-197
Collins, Francis S., 44
Comet Power Hammer, 29,254-257,**254-255**
Common Sense Power Hammer, 29,**201**
Condie, John, 20
Corr, Columbus W., **86-87**
Cortland Foundry & Machine Co., **196**-197
Cortland Machine & Tool Co., 75-76
crank-actuated hammers, (term defined) 5-**6**
Cray Brothers, 28-29
Cretor, Albert C., 199
Crucible Steel Co., 25
Cuyahoga Steam Furnace Co., 17
Davidson, Frank F., 108-111,**110**
Davis, Edwin W., **206**-207
Davis, George E., 101-102, **103**,192-195,**193**
Dawes, George, 39

Dayton Power Hammer, 92,112-114,**114**
dead stroke, (term defined) 136
Dead Stroke Power Hammer, SEE **Shaw and Justice Dead Stroke Power Hammer**
Deering, Michael, 225-229,**225**, **227**, **229**,236
Delemater, Cornelius, 56-57
Delemater Iron Works, 56-57
Depew, James, 103-106,**104-106**,191
Derbyshire, William H., 130
Dienelt & Eisenhardt, **147**-148,150, 181
disc sharpeners, **104-105**,190-**191**
Dittbenner, Hermann, **198**
Dotson, Henry & Dotson Co., 177-178
double-acting, (term defined) 17
double power hammers, 78
drop hammers, **9**-10,132,134
Dudgeon, Richard, 67
Duncan Helve Hammer, 239
Dunham's Spring Beam Hammer, 216
Dupont Manufacturing Co., 151,172
Dupont, Phillipe Denery, 45,107,150-152,**151**,197,201,213,257, 268
Dupont Power Hammer, 150-152, **151**,157,280
Easy Helve Hammer, 7,106-8,**107**, 172, **173**
Eau Claire Manufacturing Co., 257
Edlund, John A., 75
Edlund Machinery Co., 74-76
Ericsson, John, 57
Eureka Manufacturing Co., 103,196
Fairbanks, E.T. & Co., 150,169
Fairbanks Power Hammer, **151**-152
Fazendin, Leonard, 177-178
Feldhus, Heinrich, **100**,179-181,**180-181**
Fisher & Norris Co., 215
Fitzgerald, John W., 248
Flint-Lomax Electric & Manufacturing Co., 202
Flint-Lomax Power Hammer, 202-**203**
Fogelsong Machine Co., 112-113
Fogelsong, Washington, 112
Forsaith, Samuel Caldwell, 216
Frantz, Peter, 272
geometry, (term defined) 18-19
Glossop, Gilbert, 161-168,**162-164**, 210
Glossop Pneumatic Power Hammer, 161-164,**162-164**
Goodrich, J.C., 246-249
Gough, A.C., **125**-126
Grainger, Alfred J., 58-**59**
Great Depression, 74,157,176,223, 265
Griffitts, T. H., **26**
Grimshaw, William D., **124**
Grinnell Manufacturing Co., 248-250
guided-ram, (term defined) 5,135
Hackney Hammer Co., 165-168
Hackney Double-standard Power Hammer, 167
Hackney Pneumatic Power Hammer, 39,**54**,135,161-168,**165-167**
Hall, John, 34
Hallberg, A.B., 178
Hansell, H.H. & S.F., 140-141
hard-facing, 22,285

Harper's Ferry, 34
Harsco Corp., 75
Hathorn Foundry & Machine Co., 93-95
Hathorn, Henry W., **93-94**,95,109, 188,247-248
Hathorn's Hard Hitting Helve Hammer, 93-95,**94**
Hawkeye Manufacturing Co., **108-111**
Hay, James, **60**,62,90,231
Helva-Hard-Hitter, 101,**102**,103, 105,194-195
Hendrie & Boltoff Manufacturing & Supply Co., 202
Hercules "Patented" Power Hammer, 186
Hilbert-Freiberg Machine Co., 181-182
Hilbert-Freiberg Power Post Hammer, 181-**182**
Hill, Clark and Co., 164
Hjorth, B.A. & Co., 265-**266**
Holt Manufacturing Co., 90-93
Holt Power Hammer, 90-93, **91**
homemade hammers, 100,**125**,135, 178-181,**179-180**
Hooker, O.V. & Son, 150,280
Hotchkiss Atmospheric Hammer, **33**,44-51,**47**,58,63,210,234
Hotchkiss, Bennet B., 43-52,**46**,**49**, 134,161,210
Howard, Leonard D., **149**,159
Howe Power Hammer, **149**-150
Howe Scale Co., **149**-150
Hubbard & Co., 73
Hughes' Atmospheric Trip Hammer, 36,**37-38**,40
Hughes, Bernard, **36**,37-40
Hull & Belden Co., 141-142
Improved Adjustable-Helve Elastic Stroke Power-Hammer, 78, **76**
Improved Hawkeye Power Hammer, **108**
J.D. Foot & Power Hammer Combined, **206**-207
J.D. Hammer Co., 206-207
J.I. Depew Disc Sharpener Co., 103-**104**
J. Scott Rawlings, 243-244
Jinks, Benjamin F., **206**-207
Jenkins & Lingle, 67,236-242
Jenkin's Cushioned Power Hammer, 236-242,**237**,251,
Jenkins, William Rhys, Jr., 236-241, **238-240**
Johansson, Johan P., **265**-266
John Russell Cutlery Co., 29,**30**,229-**230**
Johnson, Arthur, **109**
Joseph T. Ryerson Co., 205
Joy, David, 176
Justice Hammer, **16**,143
Kane & Roach, 26,144,159-161,**160**
Kane, William E., 159-161
Kauffman, J.A., **182**
Kaukauana Machine Works, 174-175
Keiper, H.B. & L.B., 185-187
Keller Brothers, SEE Bloomer Machine Works
Keller, Charles, 101,254-257,**255**
Keller, J.E., 256
Kerrihard Co., 7,67,267-271
Kerrihard, E.W., 269

Kerrihard Power Hammer,**24**,106, 179,194,213,266-271,**267-268**,**270**,272
Kick-ass Striking Hammer, 132
Kidder Power Hammer, 261-263,**262**
Kidder, R.E. Co., 261-263
Kidder, Richard E., 261-263,**262**
Kinsey, E.A., 167
kit hammers, **181**
Knowlton Manufacturing Co., 81-86, 231
Koch, Arthur Alfred, 31,88,94,123, 187-190,**189**,212,242-250, **242**,**248**,267
Koch Manufacturing Co., 245-248
Kunz, Jean, **126**
Laird & Sweeney Manufacturing Co., 275-280
Laird & Sweeney Power Hammer, 275-280,**275**,**277**
Laird, Robert W., 275-280,**276**,**278-279**
Lane, Fredrick E., 199
Lane Tool Co., 199
Leonard, W.W., **197**,
Leonhardt, Charles, 257-**258**,
Lingle, J.H., 236-242
Little Giant Co., 169-178
Little Giant Power Hammer, 7,14,29, 106-107,120,136,150,157, 169-179,**174**, 189,199,201, 205,211
Little Hercules, 13
Little Iowa Power Hammer, **250**
Little John Power Hammer, 100,**180**
Little Modern Power Hammer, 250
Little Sampson Trip Hammer, **196-**197
Long and Allstatter Adjustable Cushioned Helve Hammer, **76**,76-81
Long & Allstatter Co., 67,76-81,83, 90
Long, John M., **77**
Longworth, Daniel, 123,243-245,**244**
Longworth Power Hammer, 243-245,**244**
Lowell Machine Shop, 35
Lower, Henry C., 190-**191**
MacGowan & Finigan Foundry & Machine Company, 104, 190-**192**
MacGowan, Peter, 191-**192**
Manning, Maxwell & Moore, 88,90, 147,150,168
Marchand & Morgan, 51
Massillon Belt Power Hammer, 199-**200**
Massillon Foundry & Machine Co., 199-200
Massillon Power Shovel Co., 200
Mayer Brothers, 7,104,106-108,**107**, 136,150,169-178,**170-173**, 188,191,196-197
Mayer, Conrad, 169
Mayer, Lorenz, 169
Mayer, Lorenz, Jr., 169-178
Mayer, Louis, 107,169-178,**171**,**175**
Mayer Power Hammer, **1**,(SEE also Little Giant), 173
McKiernan-Terry Corp., 129-130
McLane, Jared Brown, 182-185,**183-184**
McLane Power Hammer, 182-187, **183**
Mechanical Hammer Co., 127

Merrick & Towne, 4,35
Merrill Upright Power Trip Hammer, 50
Minneapolis Iron Store Co., 198-199
Minneapolis Iron Store Giant, 197-199
Miami Valley Machine Tool Co., 113
Modern Power Hammer, 94,106,110, 187,205,212,242-250,**243**, **247**,**249-250**,267
Modern Sales Co., 249-250
Moloch Foundry & Machine Co., **1**, 157,
Moloch Power Hammer, 175-176
Monarch Machine Tool Co., 75
Monson, John "Ken," 264-265
Monson, Peter, 264-265
moveable-fulcrum hammers, 88-90, 123,242-245
Murco Power Hammer, 170,176-**177**
Murray, D.J. Co., 170,176-177
Murray Power Hammer, 176-177
Nasmyth, James, 4,35-36,138
Neckyoke Singletree Hammer, 172, **173**,203
New Boss Trip Hammer, 195
New Idea Power Hammer, 196 (note number 111, page 300)
New Little Giant Trip Hammer, 170
New Metal Tool Steel Co., 274-275
New Modern Power Hammer, 250
New Power Hammer, **273**-275
New Ulm Foundry & Machine Co., 257-259
New Ulm Power Hammer, 257-259, **258**
Norquist, John, 144-**145**
Northwestern Power Hammer, 117-120,**118**,132
Northwestern Power Hammer Co., 117-121
Northwestern Power Hammer Jr., 120-**121**
Novelty Iron Works, 102-,192-195,
Novelty Works, 38
Olivers, 12-13,20,117-118,206-207
open-die, (term defined) 7-11,18-21
Ostendorf, Leslie, 25
Overcash, John F., 118-121,**119**,123,
Overman, Fredrick, 34-35
Packard, O.L., 89
Palmer, C.J., **125**-126
Palmer, Joseph, 214-216,**215**,217,273
Palmer's Power Spring Hammer, 212-216,**214**,272
parallel-bar helves, 124-132,**122-129**,
Patterson, John, **146**-147
Patterson Tool & Supply Co., 112-113
Paye, Edward, **55**,**56**,56-58,115
Payes' Hammer, **55**,
Peck, Milo, 34, 39-46,**39**,**40**,**43**,134,
Peck Manufacturing Co., 42-43
Penney's Spring Cushioned Helve Hammer, **88-89**,90,245
Penney, H.E., 89
Penney, Issac, 88-90,**89**,245
Perkins, Jacob B., 161-165
Perfect Power Hammer, 104,**105**, 190-192,**191-192**,199
Pettingel Co., 148
Philip S. Justice Co., 7,16,137,140, 225
plowshares, sharpening, 21-25,285

post hammers, (SEE bracket)
Pratt & Whitney, 51,58
Precision Castings Co., 75
Prosser, Treat T., **59**,59-60
Quickwork Co., 148
Radial Power Hammer Co., 203-206
Raymond, Edward A., 47-**48**
Raymond, Frank, 228-**229**
Reliance Engineering & Equiptment Co., 252-253
Rhodes, C.E., **179**
Rochester Helve Hammer, 73,95-97, **96**,205,
Rochester Iron Works, 37
roll-forging, **10**
rotary swaging, 10-11
rubber as engineering material, 38,65-66
Ryerson Radial Power Hammer, **204**-205
S.C. Forsaith & Co., 213-216
San Angelo Foundry and Machine Co., 189-**190**
Sarver, Grant, 132
Scranton & Co., **142**-143
Scranton Improved Upright Power Hammer, **142**-143,191
self-contained air (pneumatic) hammers, 4,13,163
Schmitt, F.J. & Co., 260
Schultz, Fritz A., 116-**117**
Schuyler Co., 201-202
Schuyler, John B. & Frank B., 202
Shaw & Justice Dead Stroke Power Hammer, 58,63,67,89,**133**, 136-143,146,149,155,**209**, 211,214-215
Shaw, Thomas, 16,45,134-141,**138-140**,143,148,212
Shaw's Gunpowder Pile-Driving Hammer, 138-**139**
Shaw's Gunpowder Forging Hammer, 138,**140**
Siebert, James D., **241**
single-acting, (term defined) 17
Southwark Foundry, 141
Spaulding, Henry W., 246-248
Square-Stroke Power Hammer, 236-241,**242**
St. Lawrence, Henry, **149**-150
Standard Power Forging Hammer, 202-**203**
Star Arms Co., 46
Star Foundry & Machine Co., 104, **105**,188-191,**189**
Star Power Hammer, **105**,120,**188**, **190**
Starite Power Hammer, 189
steam hammers, 4, 11,14,17,43, 51, 67,135,168,205,213
Sterling Hammer, **271**-272
Sterling Machine Works, 272
Strite, George T., 108-111, **110**
Suedmeier, Harlan "Sid," 178
Sutherland, Fred E., 204-**205**
Sweeney,James B., 275-280,**276**,**278-279**
Talley, B. F., **8**
Terhune, Howard, 132
Terni Steel Works, 17
Thompson, Elihu, 284
Thor Power Hammer, **101**,256
tires, welding, 22, 25-28,120,159, 172,212,284

tire-shrinkers, 25-26
trip-actuated hammers, 4,**15**(defined), **30**,35-38,58,82-83
Turpin, H.R., 194
United Hammer Co., 152
Universal Radial Power Hammer, 203-205,**204**
Up-to-date Power Hammer, 159-161,**160**
upsetters & upsetting, 10-11,172-**173**,203-204
Vaughn, Charles A., **259-261**,263-264
Victor Foot Hammer, 197
Vulcan Iron Works, 94,188
Vulcan Power Hammer (Koch), **187**-188,248
Vulcan Power Hammer (Jenkins), 236-242,**237**
Vulcanus Spring Power Hammer, 265-**266**
W.P. Duncan & Co., 236-240
Wallin, Anders, **253**-254
Weise, J.M., 194
Welch Hammer & Machine Co., 114-116
Welch Variable Blow Power Hammer, 114-**116**,124,162
Welch, William H., 114-116,**115**
welding, 11,24,27,212,283-286
West, Jonathan Burns, 97
West Tire Setter Co., 73,95-97
Western Iron & Machine Works, **181**
Williams, White & Co., 16,26,**27**, 143-146,**144-145**,158
Winters, F.R., 107
Wood, C.H., 207
Yeakley, Melvin, 163
Yoder Co., **148**
Zell & Shenk, 13